A first course in linear algebra

A FIRST COURSE IN

linear algebra

WITH CONCURRENT EXAMPLES

A.G.HAMILTON
Department of Computing Science, University of Stirling

The right of the University of Cambridge to print and sell all manner of books was granted by Henry VIII in 1534. The University has printed and published continuously since 1584.

CAMBRIDGE UNIVERSITY PRESS
Cambridge
New York New Rochelle Melbourne Sydney

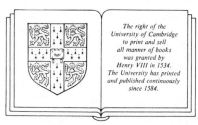

Published by the Press Syndicate of the University of Cambridge
The Pitt Building, Trumpington Street, Cambridge CB2 1RP
32 East 57th Street, New York, NY 10022, USA
10 Stamford Road, Oakleigh, Melbourne 3166, Australia

First published 1987

Printed in Great Britain

British Library cataloguing in publication data
Hamilton, A. G. (Alan G.)A first course in linear algebra: with concurrent examples.
1. Algebras, Linear
I. Title
512′.5 QA184

Library of Congress cataloging in publication data
Hamilton, A. G., 1943–
A first course in linear algebra, with concurrent examples.
Includes index.
1. Algebras, Linear. I. Title.
QA184.H36 1987 512′.55 86-24426

ISBN 0 521 32516 1 hard covers
ISBN 0 521 31041 5 paperback

MP

CONTENTS

PREFACE

Learning is not easy (not for most people, anyway). It is, of course, aided by being taught, but it is by no means only a passive exercise. One who hopes to learn must work at it actively. My intention in writing this book is not to teach, but rather to provide a stimulus and a medium through which a reader can learn. There are various sorts of textbooks with widely differing approaches. There is the encyclopaedic sort, which tends to be unreadable but contains all of the information relevant to its subject. And at the other extreme there is the work-book, which leads the reader through a progressive series of exercises. In the field of linear algebra there are already enough books of the former kind, so this book is aimed away from that end of the spectrum. But it is not a work-book, neither is it comprehensive. It is a book to be worked through, however. It is intended to be read, not referred to.

Of course, in a subject such as this, reading is not enough. Doing is also necessary. And doing is one of the main emphases of the book. It is about methods and their application. There are three aspects of this provided by this book: description, worked examples and exercises. All three are important, but I would stress that the most important of these is the exercises. In mathematics you do not know something until you can do it.

The format of the book perhaps requires some explanation. The worked examples are integrated with the text, and the careful reader will follow the examples through at the same time as reading the descriptive material. To facilitate this, the text appears on the right-hand pages only, and the examples on the left-hand pages. Thus the text and corresponding examples are visible simultaneously, with neither interrupting the other. Each chapter concludes with a set of exercises covering specifically the material of that chapter. At the end of the book there is a set of sample examination questions covering the material of the whole book.

The prerequisites required for reading this book are few. It is an introduction to the subject, and so requires only experience with methods of arithmetic, simple algebra and basic geometry. It deliberately avoids mathematical sophistication, but it presents the basis of the subject in a way which can be built on subsequently, either with a view to applications or with the development of the abstract ideas as the principal consideration.

Examples

1.1 Simple elimination (two equations).

$$2x+3y=1$$
$$x-2y=4.$$

Eliminate x as follows. Multiply the second equation by 2:

$$2x+3y=1$$
$$2x-4y=8.$$

Now replace the second equation by the equation obtained by subtracting the first equation from the second:

$$2x+3y=1$$
$$-7y=7.$$

Solve the second equation for y, giving $y=-1$. Substitute this into the first equation:

$$2x-3=1,$$

which yields $x=2$. Solution: $x=2$, $y=-1$.

1.2 Simple elimination (three equations).

$$x-2y+z=5$$
$$3x+y-z=0$$
$$x+3y+2z=2.$$

Eliminate z from the first two equations by adding them:

$$4x-y=5.$$

Next eliminate z from the second and third equations by adding twice the second to the third:

$$7x+5y=2.$$

Now solve the two simultaneous equations:

$$4x-y=5$$
$$7x+5y=2$$

as in Example 1.1. One way is to add five times the first to the second, obtaining

$$27x=27,$$

so that $x=1$. Substitute this into one of the set of two equations above which involve only x and y, to obtain (say)

$$4-y=5,$$

so that $y=-1$. Last, substitute $x=1$ and $y=-1$ into one of the original equations, obtaining

$$1+2+z=5,$$

so that $z=2$. Solution: $x=1$, $y=-1$, $z=2$.

1

Gaussian elimination

We shall describe a standard procedure which can be used to solve sets of simultaneous linear equations, no matter how many equations. Let us make sure of what the words mean before we start, however. A *linear* equation is an equation involving unknowns called x or y or z, or x_1 or x_2 or x_3, or some similar labels, in which the unknowns all occur to the first degree, which means that no squares or cubes or higher powers, and no products of two or more unknowns, occur. To *solve* a set of simultaneous equations is to find all values or sets of values for the unknowns which satisfy the equations.

Given two linear equations in unknowns x and y, as in Example 1.1, the way to proceed is to *eliminate* one of the unknowns by combining the two equations in the manner shown.

Given three linear equations in three unknowns, as in Example 1.2, we must proceed in stages. First eliminate one of the unknowns by combining two of the equations, then similarly eliminate the same unknown from a different pair of the equations by combining the third equation with one of the others. This yields two equations with two unknowns. The second stage is to solve these two equations. The third stage is to find the value of the originally eliminated unknown by substituting into one of the original equations.

This general procedure will extend to deal with n equations in n unknowns, no matter how large n is. First eliminate one of the unknowns, obtaining $n-1$ equations in $n-1$ unknowns, then eliminate another unknown from these, giving $n-2$ equations in $n-2$ unknowns, and so on until there is one equation with one unknown. Finally, substitute back to find the values of the other unknowns.

There is nothing intrinsically difficult about this procedure. It consists of the application of a small number of simple operations, used repeatedly.

1.3 The Gaussian elimination process.

$$2x_1 - x_2 + 3x_3 = 1 \qquad (1)$$
$$4x_1 + 2x_2 - x_3 = -8 \qquad (2)$$
$$3x_1 + x_2 + 2x_3 = -1 \qquad (3)$$

Stage 1:
$$x_1 - \tfrac{1}{2}x_2 + \tfrac{3}{2}x_3 = \tfrac{1}{2} \qquad (1)\div 2$$
$$4x_1 + 2x_2 - x_3 = -8 \qquad (2)$$
$$3x_1 + x_2 + 2x_3 = -1 \qquad (3)$$

Stage 2:
$$x_1 - \tfrac{1}{2}x_2 + \tfrac{3}{2}x_3 = \tfrac{1}{2} \qquad (1)$$
$$4x_2 - 7x_3 = -10 \qquad (2) - 4\times(1)$$
$$\tfrac{5}{2}x_2 - \tfrac{5}{2}x_3 = -\tfrac{5}{2} \qquad (3) - 3\times(1)$$

Stage 3:
$$x_1 - \tfrac{1}{2}x_2 + \tfrac{3}{2}x_3 = \tfrac{1}{2} \qquad (1)$$
$$x_2 - \tfrac{7}{4}x_3 = -\tfrac{5}{2} \qquad (2)\div 4$$
$$\tfrac{5}{2}x_2 - \tfrac{5}{2}x_3 = -\tfrac{5}{2} \qquad (3)$$

Stage 4:
$$x_1 - \tfrac{1}{2}x_2 + \tfrac{3}{2}x_3 = \tfrac{1}{2} \qquad (1)$$
$$x_2 - \tfrac{7}{4}x_3 = -\tfrac{5}{2} \qquad (2)$$
$$\tfrac{15}{8}x_3 = \tfrac{15}{4} \qquad (3) - \tfrac{5}{2}\times(2)$$

Stage 5:
$$x_1 - \tfrac{1}{2}x_2 + \tfrac{3}{2}x_3 = \tfrac{1}{2} \qquad (1)$$
$$x_2 - \tfrac{7}{4}x_3 = -\tfrac{5}{2} \qquad (2)$$
$$x_3 = 2. \qquad (3)\div\tfrac{15}{8}$$

Now we may obtain the solutions. Substitute $x_3 = 2$ into the second equation.

$$x_2 - \tfrac{7}{2} = -\tfrac{5}{2}, \quad \text{so } x_2 = 1.$$

Finally substitute both into the first equation, obtaining

$$x_1 - \tfrac{1}{2} + 3 = \tfrac{1}{2}, \quad \text{so } x_1 = -2.$$

Hence the solution is $x_1 = -2$, $x_2 = 1$, $x_3 = 2$.

These include multiplying an equation through by a number and adding or subtracting two equations. But, as the number of unknowns increases, the length of the procedure and the variety of different possible ways of proceeding increase dramatically. Not only this, but it may happen that our set of equations has some special nature which would cause the procedure as given above to fail: for example, a set of simultaneous equations may be *inconsistent*, i.e. have no solution at all, or, at the other end of the spectrum, it may have many different solutions. It is useful, therefore, to have a standard routine way of organising the elimination process which will apply for large sets of equations just as for small, and which will cope in a more or less automatic way with special situations. This is necessary, in any case, for the solution of simultaneous equations using a computer. Computers can handle very large sets of simultaneous equations, but they need a routine process which can be applied automatically. One such process, which will be used throughout this book, is called *Gaussian elimination*. The best way to learn how it works is to follow through examples, so Example 1.3 illustrates the stages described below, and the descriptions should be read in conjunction with it.

Stage 1 Divide the first equation through by the coefficient of x_1. (If this coefficient happens to be zero then choose another of the equations and place it first.)

Stage 2 Eliminate x_1 from the second equation by subtracting a multiple of the first equation from the second equation. Eliminate x_1 from the third equation by subtracting a multiple of the *first* equation from the third equation.

Stage 3 Divide the second equation through by the coefficient of x_2. (If this coefficient is zero then interchange the second and third equations. We shall see later how to proceed if neither of the second and third equations contains a term in x_2.)

Stage 4 Eliminate x_2 from the third equation by subtracting a multiple of the second equation.

Stage 5 Divide the third equation through by the coefficient of x_3. (We shall see later how to cope if this coefficient happens to be zero.)

At this point we have completed the elimination process. What we have done is to find another set of simultaneous equations which have the same solutions as the given set, and whose solutions can be read off very easily. What remains to be done is the following.

Read off the value of x_3. Substitute this value in the second equation, giving the value of x_2. Substitute both values in the first equation, to obtain the value of x_1.

1.4 Using arrays, solve the simultaneous equations:

$$\begin{aligned} x_1+x_2-\ x_3&=4\\ 2x_1-x_2+3x_3&=7\\ 4x_1+x_2+\ x_3&=15. \end{aligned}$$

First start with the array of coefficients:

$$\begin{array}{rrrrl}
1 & 1 & -1 & 4 & \\
2 & -1 & 3 & 7 & \\
4 & 1 & 1 & 15 & \\
\hline
1 & 1 & -1 & 4 & \\
0 & -3 & 5 & -1 & (2)-2\times(1)\\
0 & -3 & 5 & -1 & (3)-4\times(1)\\
\hline
1 & 1 & -1 & 4 & \\
0 & 1 & -\frac{5}{3} & \frac{1}{3} & (2)\div-3\\
0 & -3 & 5 & -1 & \\
\hline
1 & 1 & -1 & 4 & \\
0 & 1 & -\frac{5}{3} & \frac{1}{3} & \\
0 & 0 & 0 & 0 & (3)+3\times(2)\\
\hline
\end{array}$$

See Chapter 2 for discussion of how solutions are obtained from here.

1.5 Using arrays, solve the simultaneous equations:

$$\begin{aligned} 3x_1-3x_2+\ x_3&=1\\ -x_1+\ x_2+2x_3&=2\\ 2x_1+\ x_2-3x_3&=0. \end{aligned}$$

What follows is a full solution.

$$\begin{array}{rrrrl}
3 & -3 & 1 & 1 & \\
-1 & 1 & 2 & 2 & \\
2 & 1 & -3 & 0 & \\
\hline
1 & -1 & \frac{1}{3} & \frac{1}{3} & (1)\div 3\\
-1 & 1 & 2 & 2 & \\
2 & 1 & -3 & 0 & \\
\hline
1 & -1 & \frac{1}{3} & \frac{1}{3} & \\
0 & 0 & \frac{7}{3} & \frac{7}{3} & (2)+(1)\\
0 & 3 & -\frac{11}{3} & -\frac{2}{3} & (3)-2\times(1)\\
\hline
1 & -1 & \frac{1}{3} & \frac{1}{3} & \\
0 & 3 & -\frac{11}{3} & -\frac{2}{3} & \rbrace\ \text{interchange rows}\\
0 & 0 & \frac{7}{3} & \frac{7}{3} & \\
\hline
\end{array}$$

Notice that after stage 1 the first equation is not changed, and that after stage 3 the second equation is not changed. This is a feature of the process, however many equations there are. We proceed downwards and eventually each equation is fixed in a new form.

Besides the benefit of standardisation, there is another benefit which can be derived from this process, and that is brevity. Our working of Example 1.3 includes much that is not essential to the process. In particular the repeated writing of equations is unnecessary. Our standard process can be developed so as to avoid this, and all of the examples after Example 1.3 show the different form. The sets of equations are represented by arrays of coefficients, suppressing the unknowns and the equality signs. The first step in Example 1.4 shows how this is done. Our operations on equations now become operations on the rows of the array. These are of the following kinds:

- interchange rows,
- divide (or multiply) one row through by a number,
- subtract (or add) a multiple of one row from (to) another.

These are called *elementary row operations*, and they play a large part in our later work. It is important to notice the form of the array at the end of the process. It has a triangle of 0s in the lower left corner and 1s down the diagonal from the top left.

Now let us take up two complications mentioned above. In stage 5 of the Gaussian elimination process (henceforward called the GE process) the situation not covered was when the coefficient of x_3 in the third equation (row) was zero. In this case we divide the third equation (row) by the number occurring on the right-hand side (in the last column), if this is not already zero. Example 1.4 illustrates this. The solution of sets of equations for which this happens will be discussed in the next chapter. What happens is that either the equations have no solutions or they have infinitely many solutions.

The other complication can arise in stage 3 of the GE process. Here the coefficient of x_2 may be zero. The instruction was to interchange equations (rows) in the hope of placing a non-zero coefficient in this position. When working by hand we may choose which row to interchange with so as to make the calculation easiest (presuming that there is a choice). An obvious way to do this is to choose a row in which this coefficient is 1. Example 1.5 shows this being done. When the GE process is formalised (say for computer application), however, we need a more definite rule, and the one normally adopted is called *partial pivoting*. Under this rule, when we interchange rows because of a zero coefficient, we choose to interchange with the row which has the coefficient which is numerically the *largest* (that

1	−1	$\frac{1}{3}$	$\frac{1}{3}$	
0	1	$-\frac{11}{9}$	$-\frac{2}{9}$	(2)÷3
0	0	1	1	(3)÷$\frac{7}{3}$

From here, $x_3=1$, and substituting back we obtain

$$x_2-\tfrac{11}{9}=-\tfrac{2}{9},\quad \text{so } x_2=1.$$

Substituting again:

$$x_1-1+\tfrac{1}{3}=\tfrac{1}{3},\quad \text{so } x_1=1.$$

Hence the solution sought is: $x_1=1$, $x_2=1$, $x_3=1$.

1.6 Using arrays, solve the simultaneous equations:

$$\begin{aligned} x_1+x_2-x_3&=-3\\ 2x_1+2x_2+x_3&=0\\ 5x_1+5x_2-3x_3&=-8. \end{aligned}$$

Solution:

1	1	−1	−3	
2	2	1	0	
5	5	−3	−8	
1	1	−1	−3	
0	0	3	6	(2)−2×(1)
0	0	2	7	(3)−5×(1)
1	1	−1	−3	
0	0	1	2	(2)÷3
0	0	2	7	
1	1	−1	−3	
0	0	1	2	
0	0	0	3	(3)−2×(2)

Next, and finally, divide the last row by 3. How to obtain solutions from this point is discussed in Chapter 2. (In fact there are no solutions in this case.)

1.7 Solve the simultaneous equations:

$$\begin{aligned} 2x_1-2x_2+x_3-3x_4&=2\\ x_1-x_2+3x_3-x_4&=-2\\ -x_1-2x_2+x_3+2x_4&=-6\\ 3x_1+x_2-x_3-2x_4&=7. \end{aligned}$$

Convert to an array and proceed:

2	−2	1	−3	2
1	−1	3	−1	−2
−1	−2	1	2	−6
3	1	−1	−2	7

is, the largest when any negative signs are disregarded). This has two benefits. First, we (and more particularly, the computer) know precisely what to do at each stage and, second, following this process actually produces a more accurate answer when calculations are subject to rounding errors, as will always be the case with computers. Generally, we shall not use partial pivoting, since our calculations will all be done by hand with small-scale examples.

There may be a different problem at stage 3. We may find that there is no equation (row) which we can choose which has a non-zero coefficient in the appropriate place. In this case we do nothing, and just move on to consideration of x_3, as shown in Example 1.6. How to solve the equations in such a case is discussed in the next chapter.

The GE process has been described above in terms which can be extended to cover larger sets of equations (and correspondingly larger arrays of coefficients). We should bear in mind always that the form of the array which we are seeking has rows in which the first non-zero coefficient (if there is one) is 1, and this 1 is to the *right* of the first non-zero coefficient in the preceding row. Such a form for an array is called *row-echelon form.* Example 1.7 shows the process applied to a set of four equations in four unknowns.

Further examples of the GE process applied to arrays are given in the following exercises. Of course the way to learn this process is to carry it out, and the reader is recommended not to proceed to the rest of the book before gaining confidence in applying it.

Summary

The purpose of this chapter is to describe the Gaussian elimination process which is used in the solution of simultaneous equations, and the abbreviated way of carrying it out, using elementary row operations on rectangular arrays.

$$\begin{array}{rrrrrl}
1 & -1 & \frac{1}{2} & -\frac{3}{2} & 1 & (1)\div 2\\
1 & -1 & 3 & -1 & -2 & \\
-1 & -2 & 1 & 2 & -6 & \\
3 & 1 & -1 & -2 & 7 & \\
\hline
1 & -1 & \frac{1}{2} & -\frac{3}{2} & 1 & \\
0 & 0 & \frac{5}{2} & \frac{1}{2} & -3 & (2)-(1)\\
0 & -3 & \frac{3}{2} & \frac{1}{2} & -5 & (3)+(1)\\
0 & 4 & -\frac{5}{2} & \frac{5}{2} & 4 & (4)-3\times(1)\\
\hline
1 & -1 & \frac{1}{2} & -\frac{3}{2} & 1 & \\
0 & -3 & \frac{3}{2} & \frac{1}{2} & -5 & \rbrace\ \text{interchange rows}\\
0 & 0 & \frac{5}{2} & \frac{1}{2} & -3 & \\
0 & 4 & -\frac{5}{2} & \frac{5}{2} & 4 & \\
\hline
1 & -1 & \frac{1}{2} & -\frac{3}{2} & 1 & \\
0 & 1 & -\frac{1}{2} & -\frac{1}{6} & \frac{5}{3} & (2)\div -3\\
0 & 0 & \frac{5}{2} & \frac{1}{2} & -3 & \\
0 & 4 & -\frac{5}{2} & \frac{5}{2} & 4 & \\
\hline
1 & -1 & \frac{1}{2} & -\frac{3}{2} & 1 & \\
0 & 1 & -\frac{1}{2} & -\frac{1}{6} & \frac{5}{3} & \\
0 & 0 & \frac{5}{2} & \frac{1}{2} & -3 & \\
0 & 0 & -\frac{1}{2} & \frac{19}{6} & -\frac{8}{3} & (4)-4\times(2)\\
\hline
1 & -1 & \frac{1}{2} & -\frac{3}{2} & 1 & \\
0 & 1 & -\frac{1}{2} & -\frac{1}{6} & \frac{5}{3} & \\
0 & 0 & 1 & \frac{1}{5} & -\frac{6}{5} & (3)\div \frac{5}{2}\\
0 & 0 & -\frac{1}{2} & \frac{19}{6} & -\frac{8}{3} & \\
\hline
1 & -1 & \frac{1}{2} & -\frac{3}{2} & 1 & \\
0 & 1 & -\frac{1}{2} & -\frac{1}{6} & \frac{5}{3} & \\
0 & 0 & 1 & \frac{1}{5} & -\frac{6}{5} & \\
0 & 0 & 0 & \frac{49}{15} & -\frac{49}{15} & (4)+\frac{1}{2}\times(3)\\
\hline
1 & -1 & \frac{1}{2} & -\frac{3}{2} & 1 & \\
0 & 1 & -\frac{1}{2} & -\frac{1}{6} & \frac{5}{3} & \\
0 & 0 & 1 & \frac{1}{5} & -\frac{6}{5} & \\
0 & 0 & 0 & 1 & -1 & (4)\div \frac{49}{15}\\
\hline
\end{array}$$

From this last row we deduce that $x_4 = -1$. Substituting back gives:

$$x_3 - \tfrac{1}{5} = -\tfrac{6}{5}, \quad \text{so } x_3 = -1,$$
$$x_2 + \tfrac{1}{2} + \tfrac{1}{6} = \tfrac{5}{3}, \quad \text{so } x_2 = 1, \quad \text{and}$$
$$x_1 - 1 - \tfrac{1}{2} + \tfrac{3}{2} = 1, \quad \text{so } x_1 = 1.$$

Hence the solution is: $x_1 = 1$, $x_2 = 1$, $x_3 = -1$, $x_4 = -1$.

Exercises

Using the Gaussian elimination process, solve the following sets of simultaneous equations.

(i)
$$\begin{aligned} x-y&=2\\ 2x+y&=1. \end{aligned}$$

(ii)
$$\begin{aligned} 3x+2y&=0\\ x-y&=5. \end{aligned}$$

(iii)
$$\begin{aligned} x_1+x_2+x_3&=2\\ 2x_1+x_2-2x_3&=0\\ -x_1-2x_2+3x_3&=4. \end{aligned}$$

(iv)
$$\begin{aligned} 3x_1-x_2-x_3&=7\\ x_1-x_2+x_3&=0\\ -x_1+2x_2+2x_3&=-2. \end{aligned}$$

(v)
$$\begin{aligned} 2x_1-4x_2+x_3&=2\\ x_1-2x_2-2x_3&=-4\\ -x_1+x_2&=-1. \end{aligned}$$

(vi)
$$\begin{aligned} -x_1+2x_2-x_3&=-2\\ 4x_1-x_2-2x_3&=2\\ 3x_1-4x_3&=-1. \end{aligned}$$

(vii)
$$\begin{aligned} 2x_2-x_3&=-5\\ 3x_1-x_2+2x_3&=8\\ x_1+2x_2+2x_3&=5. \end{aligned}$$

(viii)
$$\begin{aligned} x_1+5x_2-2x_3&=0\\ 3x_1-x_2+10x_3&=0\\ -x_1-2x_2+7x_3&=0. \end{aligned}$$

(ix)
$$\begin{aligned} x_1-3x_2-x_3&=0\\ 2x_1-4x_2-7x_3&=2\\ 7x_1-13x_2&=8 \end{aligned}$$

(x)
$$\begin{aligned} -2x_1-x_2-x_3&=-2\\ 3x_2-7x_3&=0\\ -3x_1+x_2-4x_3&=3 \end{aligned}$$

(xi)
$$\begin{aligned} x_1-2x_2+x_3-x_4&=1\\ -x_1-x_2+2x_3+2x_4&=-5\\ 2x_1+x_2+x_3+3x_4&=0\\ x_1+3x_2-3x_3+3x_4&=2. \end{aligned}$$

(xii)
$$\begin{aligned} x_1+x_2-3x_3+x_4&=-2\\ 2x_1+2x_2+x_3+3x_4&=0\\ x_1+2x_2-2x_3+2x_4&=-2\\ -3x_2+4x_3-x_4&=1. \end{aligned}$$

Examples

2.1 Find all values of x and y which satisfy the equations:

$$x+2y=1$$
$$3x+6y=3.$$

GE process:

$$\begin{array}{ccc} 1 & 2 & 1 \\ 3 & 6 & 3 \\ \hline 1 & 2 & 1 \\ 0 & 0 & 0 \\ \hline \end{array} \qquad (2)-3\times(1)$$

Here the second row gives no information. All we have to work with is the single equation $x+2y=1$. Set $y=t$ (say). Then on substitution we obtain $x=1-2t$. Hence all values of x and y which satisfy the equations are given by:

$$x=1-2t, \quad y=t \quad (t\in\mathbb{R}).$$

2.2 Find all solutions to:

$$x_1+x_2-\ x_3=\ 5$$
$$3x_1-x_2+2x_3=-2.$$

The GE process yields:

$$\begin{array}{cccc} 1 & 1 & -1 & 5 \\ 0 & 1 & -\frac{5}{4} & \frac{17}{4} \end{array}$$

Set $x_3=t$. Substituting then gives

$$x_2-\tfrac{5}{4}t=\tfrac{17}{4}, \quad \text{so } x_2=\tfrac{17}{4}+\tfrac{5}{4}t, \quad \text{and}$$
$$x_1+(\tfrac{17}{4}+\tfrac{5}{4}t)-t=5, \quad \text{so } x_1=\tfrac{3}{4}+\tfrac{1}{4}t.$$

Hence the full solution is:

$$x_1=\tfrac{3}{4}+\tfrac{1}{4}t, \quad x_2=\tfrac{17}{4}+\tfrac{5}{4}t, \quad x_3=t \quad (t\in\mathbb{R}).$$

2.3 Solve the equations:

$$x_1-\ x_2-4x_3=0$$
$$3x_1+\ x_2-\ x_3=3$$
$$5x_1+3x_2+2x_3=6.$$

The GE process yields:

$$\begin{array}{cccc} 1 & -1 & -4 & 0 \\ 0 & 1 & \frac{11}{4} & \frac{3}{4} \\ 0 & 0 & 0 & 0 \end{array}$$

In effect, then, we have only two equations to solve for three unknowns. Set $x_3=t$. Substituting then gives

$$x_2+\tfrac{11}{4}t=\tfrac{3}{4}, \quad \text{so } x_2=\tfrac{3}{4}-\tfrac{11}{4}t, \quad \text{and}$$
$$x_1-(\tfrac{3}{4}-\tfrac{11}{4}t)-4t=0, \quad \text{so } x_1=\tfrac{3}{4}-\tfrac{5}{4}t.$$

Hence the full solution is: $x_1=\frac{3}{4}-\frac{5}{4}t$, $x_2=\frac{3}{4}-\frac{11}{4}t$, $x_3=t$ $(t\in\mathbb{R})$.

2

Solutions to simultaneous equations 1

Now that we have a routine procedure for the elimination of variables (Gaussian elimination), we must look more closely at where it can lead, and at the different possibilities which can arise when we seek solutions to given simultaneous equations.

Example 2.1 illustrates in a simple way one possible outcome. After the GE process the second row consists entirely of zeros and is thus of no help in finding solutions. This has happened because the original second equation is a multiple of the first equation, so in essence we are given only a single equation connecting the two variables. In such a situation there are infinitely many possible solutions. This is because we may specify *any* value for one of the unknowns (say y) and then the equation will give the value of the other unknown. Thus the customary form of the solution to Example 2.1 is:

$$y=t, \quad x=1-2t \quad (t \in \mathbb{R}).$$

These ideas extend to the situation generally when there are fewer equations than unknowns. Example 2.2 illustrates the case of two equations with three unknowns. We may specify any value for one of the unknowns (here put $z=t$) and then solve the two equations for the other two unknowns. This situation may also arise when we are originally given three equations in three unknowns, as in Example 2.3. See also Example 1.4.

2.4 Find all solutions to the set of equations:

$$x_1+x_2+x_3=1$$
$$x_1+x_2+x_3=4.$$

This is a nonsensical problem. There are no values of x_1, x_2 and x_3 which satisfy both of these equations simultaneously. What does the GE process give us here?

1	1	1	1	
1	1	1	4	
1	1	1	1	
0	0	0	3	(2) − (1)
1	1	1	1	
0	0	0	1	(2) ÷ 3

The last row, when transformed back into an equation, is

$$0x_1+0x_2+0x_3=1.$$

This is satisfied by *no* values of x_1, x_2 and x_3.

2.5 Find all solutions to the set of equations:

$$x_1+2x_2+\ x_3=1$$
$$2x_1+5x_2-\ x_3=4$$
$$x_1+\ x_2+4x_3=2.$$

GE process:

1	2	1	1	
2	5	−1	4	
1	1	4	2	
1	2	1	1	
0	1	−3	2	(2) − 2 × (1)
0	−1	3	1	(3) − (1)
1	2	1	1	
0	1	−3	2	
0	0	0	3	(3) + (2)
1	2	1	1	
0	1	−3	2	
0	0	0	1	(3) ÷ (3)

Because of the form of this last row, we can say straight away that there are no solutions in this case (indeed, the last step was unnecessary: a last row of 0 0 0 3 indicates inconsistency immediately).

Here, then, is a simple-minded rule: if there are fewer equations than unknowns then there will be infinitely many solutions (if there are solutions at all). This rule is more usefully applied *after* the GE process has been completed, because the original equations may disguise the real situation, as in Examples 2.1, 2.3 and 1.4.

The qualification must be placed on this rule because such sets of equations may have no solutions at all. Example 2.4 is a case in point. Two equations, three unknowns, and no solutions. These equations are clearly *inconsistent* equations. There are no values of the unknowns which satisfy both. In such a case it is obvious that they are inconsistent. The equations in Example 2.5 are also inconsistent, but it is not obvious there. The GE process automatically tells us when equations are inconsistent. In Example 2.5 the last row turns out to be

$$0 \quad 0 \quad 0 \quad 1,$$

which, if translated back into an equation, gives

$$0x_1 + 0x_2 + 0x_3 = 1,$$

i.e.

$$0 = 1.$$

When this happens, the conclusion that we can draw is that the given equations are inconsistent and have no solutions. See also Example 1.6. This may happen whether there are as many equations as unknowns, more equations, or fewer equations.

2.6 Find all solutions to the set of equations:

$$x_1 + x_2 = 2$$
$$3x_1 - x_2 = 2$$
$$-x_1 + 2x_2 = 3.$$

GE process:

1	1	2	
3	−1	2	
−1	2	3	
1	1	2	
0	−4	−4	(2) − 3 × (1)
0	3	5	(3) + (1)
1	1	2	
0	1	1	(2) ÷ −4
0	3	5	
1	1	2	
0	1	1	
0	0	2	(3) − 3 × (2)

Without performing the last step of the standard process we can see here that the given equations are inconsistent.

2.7 Find all solutions to the set of equations:

$$x_1 - 4x_2 = -1$$
$$2x_1 + 2x_2 = 8$$
$$5x_1 - x_2 = 14.$$

GE process:

1	−4	−1	
2	2	8	
5	−1	14	
1	−4	−1	
0	10	10	(2) − 2 × (1)
0	19	19	(3) − 5 × (1)
1	−4	−1	
0	1	1	(2) ÷ 10
0	0	0	(after two steps)

Here there is a solution. The third equation is in effect redundant. The second row yields $x_2 = 1$. Substituting in the first gives:

$$x_1 - 4 = -1, \quad \text{so } x_1 = 3.$$

Hence the solution is: $x_1 = 3$, $x_2 = 1$.

Example 2.6 has three equations with two unknowns. Here there are more equations than we need to determine the values of the unknowns. We can think of using the first two equations to find these values and then trying them in the third. If we are lucky they will work! But the more likely outcome is that such sets of equations are inconsistent. Too many equations may well lead to inconsistency. But not always. See Example 2.7.

We can now see that there are three possible outcomes when solving simultaneous equations:

(i) there is no solution,

(ii) there is a unique solution,

(iii) there are infinitely many solutions.

One of the most useful features of the GE process is that it tells us automatically which of these occurs, in advance of finding the solutions.

2.8 Illustration of the various possibilities arising from the GE process and the nature of the solutions indicated.

(i) $\begin{bmatrix} 1 & 2 & 1 \\ 0 & 1 & 3 \end{bmatrix}$ unique solution.

(ii) $\begin{bmatrix} 1 & -1 & 2 \\ 0 & 0 & 1 \end{bmatrix}$ inconsistent.

(iii) $\begin{bmatrix} 1 & 3 & 3 \\ 0 & 1 & 0 \end{bmatrix}$ unique solution.

(iv) $\begin{bmatrix} 1 & 1 & 2 \\ 0 & 0 & 0 \end{bmatrix}$ infinitely many solutions.

(v) $\begin{bmatrix} 1 & 2 & 1 & 4 \\ 0 & 1 & -2 & 2 \\ 0 & 0 & 1 & 3 \end{bmatrix}$ unique solution.

(vi) $\begin{bmatrix} 1 & 0 & -1 & 5 \\ 0 & 1 & 1 & -3 \\ 0 & 0 & 0 & 1 \end{bmatrix}$ inconsistent.

(vii) $\begin{bmatrix} 1 & 3 & 0 & 2 \\ 0 & 1 & 3 & -1 \\ 0 & 0 & 1 & 0 \end{bmatrix}$ unique solution.

(viii) $\begin{bmatrix} 1 & -1 & 1 & 5 \\ 0 & 1 & 7 & 2 \\ 0 & 0 & 0 & 0 \end{bmatrix}$ infinitely many solutions.

(ix) $\begin{bmatrix} 1 & 2 & -1 & 3 \\ 0 & 0 & 1 & 2 \\ 0 & 0 & 0 & 1 \end{bmatrix}$ inconsistent.

(x) $\begin{bmatrix} 1 & 2 & 2 & 5 \\ 0 & 0 & 1 & 2 \\ 0 & 0 & 0 & 0 \end{bmatrix}$ infinitely many solutions.

Rule

Given a set of (any number of) simultaneous equations in p unknowns:

(i) there is no solution if after the GE process the last non-zero row has a 1 at the right-hand end and zeros elsewhere;

(ii) there is a unique solution if after the GE process there are exactly p non-zero rows, the last of which has a 1 in the position *second* from the right-hand end;

(iii) there are infinitely many solutions if after the GE process there are fewer than p non-zero rows and (i) above does not apply.

Example 2.8 gives various arrays resulting from the GE process, to illustrate the three possibilities above. Note that the number of unknowns is always one fewer than the number of columns in the array.

2.9 Find all values of c for which the equations

$$x+y=c$$
$$3x-cy=2$$

have a solution.

GE process:

$$\begin{array}{ccc} 1 & 1 & c \\ 3 & -c & 2 \\ \hline 1 & 1 & c \\ 0 & -c-3 & 2-3c \\ \hline 1 & 1 & c \\ 0 & 1 & \dfrac{2-3c}{-c-3} \\ \hline \end{array}$$

(2) − 3 × (1)

(2) ÷ (−c − 3)

Now this last step is legitimate only if $-c-3$ is not zero. Thus, provided that $c+3\neq 0$, we can say

$$y=\frac{2-3c}{-c-3} \quad \text{and} \quad x=c-\frac{2-3c}{-c-3}.$$

If $c+3=0$ then $c=-3$, and the equations are

$$x+y=-3$$
$$3x+3y=2.$$

These are easily seen to be inconsistent. Thus the given equations have a solution if and only if $c\neq -3$.

2.10 Find all values of c for which the following equations have
(a) a unique solution,
(b) no solution,
(c) infinitely many solutions.

$$x_1+x_2+x_3=c$$
$$cx_1+x_2+2x_3=2$$
$$x_1+cx_2+x_3=4.$$

GE process:

$$\begin{array}{cccc} 1 & 1 & 1 & c \\ c & 1 & 2 & 2 \\ 1 & c & 1 & 4 \\ \hline 1 & 1 & 1 & c \\ 0 & 1-c & 2-c & 2-c^2 \\ 0 & c-1 & 0 & 4-c \quad * \\ \hline \end{array}$$

Finally, Examples 2.9 and 2.10 show how the GE process can be applied even when the coefficients in the given simultaneous equations involve a parameter (or parameters) which may be unknown or unspecified. As naturally expected, the solution values for the unknowns depend on the parameter(s), but, importantly, the nature of the solution, that is to say, whether there are no solutions, a unique solution or infinitely many solutions, also depends on the value(s) of the parameter(s).

Summary

This chapter should enable the reader to apply the GE process to any given set of simultaneous linear equations to find whether solutions exist, and if they do to determine whether there is a unique solution or infinitely many solutions, and to find them.

Exercises

1. Show that the following sets of equations are inconsistent.

(i) $x-2y=1$, $2x-y=-8$, $-x+y=6$.

(ii) $3x+y=3$, $2x-y=7$, $5x+4y=4$.

(iii) $x_1-x_2-2x_3=-1$, $-2x_1+x_2+x_3=2$, $3x_1+2x_2+9x_3=4$.

(iv) $2x_1-x_2+4x_3=4$, $x_1+2x_2-3x_3=1$, $3x_1+3x_3=6$.

2. Show that the following sets of equations have infinitely many solutions, and express the solutions in terms of parameters.

(i) $x-3y=2$, $2x-6y=4$.

(ii) $2x+3y=-1$, $8x+12y=-4$.

(iii) $x_1+x_2+x_3=5$, $-x_1+2x_2-7x_3=-2$, $2x_1+x_2+4x_3=9$.

(iv) $x_1+2x_3=1$, $2x_1+x_2+3x_3=1$, $2x_1-x_2+5x_3=3$.

(v) $x_1-x_2+3x_3=4$, $2x_1-2x_2+x_3=3$, $-x_1+x_2+x_3=0$.

(vi) $x_1+2x_2+x_3=2$, $2x_1-x_2+7x_3=-6$, $-x_1+x_2-4x_3=4$, $x_1-2x_2+5x_3=-6$.

3. Show that the following sets of equations have unique solutions, and find them.

(i) $2x-5y=-1$, $3x+y=7$.

(ii) $x-2y=-1$, $4x+y=14$, $3x-4y=-3$.

(iii) $x_1-x_2-2x_3=-6$, $3x_1+x_2-2x_3=-6$, $-2x_1-2x_2+x_3=2$.

(iv) $3x_2+x_3=-3$, $x_1-2x_2-2x_3=4$, $2x_1+x_2-3x_3=3$.

$$\begin{array}{cccc}
1 & 1 & 1 & c \\
0 & 1 & \dfrac{2-c}{1-c} & \dfrac{2-c^2}{1-c} \\
0 & 0 & 2-c & 4-c+2-c^2 \quad ** \\
\end{array} \qquad \text{(provided that } c \neq 1)$$

$$\begin{array}{cccc}
1 & 1 & 1 & c \\
0 & 1 & \dfrac{2-c}{1-c} & \dfrac{2-c^2}{1-c} \\
0 & 0 & 1 & 3+c \\
\end{array} \qquad \text{(provided that } c \neq 2)$$

If $c=1$ then the row marked * is 0 0 0 3, showing the equations to be inconsistent. If $c=2$ then the row marked ** is 0 0 0 0, and the equations have infinitely many solutions: $x_3=t, x_2=t, x_1=-t \ (t \in \mathbb{R})$. Last, if $c \neq 1$ and $c \neq 2$ then there is a unique solution, given by the last array above:

$$x_3 = 3+c,$$

$$x_2 = \frac{2-c^2}{1-c} - \frac{(2-c)(3+c)}{1-c},$$

and

$$x_1 = c - \frac{2-c^2}{1-c} + \frac{(2-c)(3+c)}{1-c} - (3+c).$$

(v)
$$\begin{aligned} x_1+2x_2+3x_3&=3\\ 2x_1+3x_2+4x_3&=3\\ 3x_1+4x_2+5x_3&=3\\ x_1+x_2+x_3&=0. \end{aligned}$$

(vi)
$$\begin{aligned} -x_1+x_2+x_3+x_4&=2\\ x_1-x_2+x_3+x_4&=4\\ x_1+x_2-x_3+x_4&=6\\ x_1+x_2+x_3-x_4&=8. \end{aligned}$$

4. Find whether the following sets of equations are inconsistent, have a unique solution, or have infinitely many solutions.

(i)
$$\begin{aligned} x_1+x_2+x_3&=1\\ 2x_1+x_2-3x_3&=1\\ 3x_2-x_3&=1. \end{aligned}$$

(ii)
$$\begin{aligned} x_1+2x_2-x_3&=2\\ 2x_1+2x_2-4x_3&=0\\ -x_1+3x_3&=2. \end{aligned}$$

(iii)
$$\begin{aligned} x_1-x_2+x_3-2x_4&=-6\\ 2x_2+x_3-3x_4&=-5\\ 3x_1-x_2-4x_3-x_4&=9\\ -x_1-3x_2+3x_3+2x_4&=-5. \end{aligned}$$

(iv)
$$\begin{aligned} x_1+x_2+x_3&=2\\ 3x_1-2x_2+x_3&=3\\ 2x_1+x_3&=3\\ -x_1+3x_2+x_3&=3. \end{aligned}$$

(v)
$$\begin{aligned} x_1+x_2+x_3+x_4&=0\\ x_1+x_4&=0\\ x_1+2x_2+x_3&=0. \end{aligned}$$

5. (i) Examine the solutions of

$$\begin{aligned} x_1-x_2+x_3&=c\\ 2x_1-3x_2+4x_3&=0\\ 3x_1-4x_2+5x_3&=1, \end{aligned}$$

when $c=1$ and when $c\neq 1$.

(ii) Find all values of k for which the equations

$$\begin{aligned} 3x_1-7x_2-4x_3&=8\\ -2x_1+6x_2+11x_3&=21\\ -5x_1-21x_2+7x_3&=10k\\ x_1+23x_2+13x_3&=41 \end{aligned}$$

are consistent.

6. In the following sets of equations, determine all values of c for which the set of equations has (a) no solution, (b) infinitely many solutions, and (c) a unique solution.

(i)
$$\begin{aligned} x_1+x_2-x_3&=2\\ x_1+2x_2+x_3&=3\\ x_1+x_2+(c^2-5)x_3&=c. \end{aligned}$$

(ii)
$$\begin{aligned} x_1+x_2+x_3&=2\\ 2x_1+3x_2+2x_3&=5\\ 2x_1+3x_2+(c^2-1)x_3&=c+1. \end{aligned}$$

(iii)
$$\begin{aligned} x_1+x_2&=3\\ x_1+(c^2-8)x_2&=c. \end{aligned}$$

(iv)
$$\begin{aligned} cx_1+x_2-2x_3&=0\\ x_1+cx_2+3x_3&=0\\ 2x_1+3x_2+cx_3&=0. \end{aligned}$$

Examples

3.1 Examples of matrix notations.

$$A=\begin{bmatrix} 1 & 2 & 3 \\ -2 & -1 & 0 \end{bmatrix}: \quad \begin{matrix} a_{11}=1, & a_{12}=2, & a_{13}=3, \\ a_{21}=-2, & a_{22}=-1, & a_{23}=0. \end{matrix}$$

$$B=\begin{bmatrix} 5 & 6 \\ 7 & 8 \\ 9 & 10 \end{bmatrix}. \quad \begin{matrix} b_{11}=5, & b_{12}=6, \\ b_{21}=7, & b_{22}=8, \\ b_{31}=9, & b_{32}=10. \end{matrix}$$

A is a 2×3 matrix, B is a 3×2 matrix.

3.2 Examples of matrix addition.

$$\begin{bmatrix} 1 & 2 & 3 \\ 4 & 5 & 6 \end{bmatrix}+\begin{bmatrix} 3 & 2 & 1 \\ 2 & 3 & 4 \end{bmatrix}=\begin{bmatrix} 4 & 4 & 4 \\ 6 & 8 & 10 \end{bmatrix}.$$

$$\begin{bmatrix} -1 & 3 & 2 \\ 4 & 0 & 1 \\ -2 & 1 & 5 \end{bmatrix}+\begin{bmatrix} 4 & 1 & 1 \\ 3 & 2 & -2 \\ 1 & 2 & -3 \end{bmatrix}=\begin{bmatrix} 3 & 4 & 3 \\ 7 & 2 & -1 \\ -1 & 3 & 2 \end{bmatrix}.$$

$$\begin{bmatrix} 6 & 1 \\ -1 & 2 \\ 3 & 4 \end{bmatrix}-\begin{bmatrix} 3 & 2 \\ 1 & -3 \\ 0 & 1 \end{bmatrix}=\begin{bmatrix} 3 & -1 \\ -2 & 5 \\ 3 & 3 \end{bmatrix}.$$

$$\begin{bmatrix} a_{11} & a_{12} & a_{13} \\ a_{21} & a_{22} & a_{23} \end{bmatrix}+\begin{bmatrix} b_{11} & b_{12} & b_{13} \\ b_{21} & b_{22} & b_{23} \end{bmatrix}=\begin{bmatrix} a_{11}+b_{11} & a_{12}+b_{12} & a_{13}+b_{13} \\ a_{21}+b_{21} & a_{22}+b_{22} & a_{23}+b_{23} \end{bmatrix}.$$

3.3 Examples of scalar multiples.

$$\begin{bmatrix} 5 & 6 \\ 7 & 8 \\ 9 & 10 \end{bmatrix}+\begin{bmatrix} 5 & 6 \\ 7 & 8 \\ 9 & 10 \end{bmatrix}=\begin{bmatrix} 10 & 12 \\ 14 & 16 \\ 18 & 20 \end{bmatrix}=2\begin{bmatrix} 5 & 6 \\ 7 & 8 \\ 9 & 10 \end{bmatrix}.$$

$$6\begin{bmatrix} 1 & 2 \\ 3 & 4 \end{bmatrix}=\begin{bmatrix} 6 & 12 \\ 18 & 24 \end{bmatrix}.$$

$$\tfrac{1}{2}\begin{bmatrix} 3 & -2 & 1 \\ 2 & 1 & -4 \end{bmatrix}=\begin{bmatrix} \frac{3}{2} & -1 & \frac{1}{2} \\ 1 & \frac{1}{2} & -2 \end{bmatrix}.$$

3.4 More scalar multiples.

Let $A=\begin{bmatrix} -1 & 1 \\ 2 & 4 \end{bmatrix}$ and $B=\begin{bmatrix} -1 & 3 & 2 \\ 4 & 0 & 1 \\ -2 & 1 & 5 \end{bmatrix}$.

Then

$$2A=\begin{bmatrix} -2 & 2 \\ 4 & 8 \end{bmatrix},\quad 7A=\begin{bmatrix} -7 & 7 \\ 14 & 28 \end{bmatrix},\quad \tfrac{1}{2}A=\begin{bmatrix} -\frac{1}{2} & \frac{1}{2} \\ 1 & 2 \end{bmatrix}$$

and $2B=\begin{bmatrix} -2 & 6 & 4 \\ 8 & 0 & 2 \\ -4 & 2 & 10 \end{bmatrix}$, $-5B=\begin{bmatrix} 5 & -15 & -10 \\ -20 & 0 & -5 \\ 10 & -5 & -25 \end{bmatrix}$, $\frac{1}{5}B=\begin{bmatrix} -\frac{1}{5} & \frac{3}{5} & \frac{2}{5} \\ \frac{4}{5} & 0 & \frac{1}{5} \\ -\frac{2}{5} & \frac{1}{5} & 1 \end{bmatrix}$.

3

Matrices and algebraic vectors

A *matrix* is nothing more than a rectangular array of numbers (for us this means real numbers). In fact the arrays which were part of the shorthand way of carrying out the Gaussian elimination process are matrices. The usefulness of matrices originates in precisely that process, but extends far beyond. We shall see in this chapter how the advantages of brevity gained through the use of arrays in Chapters 1 and 2 can be developed, and how out of this development the idea of a matrix begins to stand on its own.

An array of numbers with p rows and q columns is called a $p \times q$ matrix ('p by q matrix'), and the numbers themselves are called the *entries* in the matrix. The number in the ith row and jth column is called the (i, j)-entry. Sometimes suffixes are used to indicate position, so that a_{ij} (or b_{ij}, etc.) may be used for the (i, j)-entry. The first suffix denotes the row and the second suffix the column. See Examples 3.1. A further notation which is sometimes used is $[a_{ij}]_{p \times q}$. This denotes the $p \times q$ matrix whose (i, j)-entry is a_{ij}, for each i and j.

Immediately we can see that there are extremes allowed under this definition, namely when either p or q is 1. When p is 1 the matrix has only one row, and is called a *row vector*, and when q is 1 the matrix has only one column, and is called a *column vector*. The case when both p and q are 1 is rather trivial and need not concern us here. A column vector with p entries we shall call a *p-vector*, so a p-vector is a $p \times 1$ matrix.

Addition of matrices (including addition of row or column vectors) is very straightforward. We just add the corresponding entries. See Examples 3.2. The only point to note is that, in order for the sum of two matrices (or vectors) to make sense, they must be of the same size. To put this precisely, they must both be $p \times q$ matrices, for the same p and q. In formal terms, if A is the $p \times q$ matrix whose (i, j)-entry is a_{ij} and B is the $p \times q$ matrix whose (i, j)-entry is b_{ij} then $A + B$ is the $p \times q$ matrix whose (i, j)-entry is $a_{ij} + b_{ij}$. Likewise subtraction: $A - B$ is the $p \times q$ matrix whose (i, j)-entry is $a_{ij} - b_{ij}$.

3.5 Multiplication of a matrix with a column vector. Consider the equations:

$$\begin{aligned} 2x_1 - x_2 + 4x_3 &= 1 \\ x_1 + 3x_2 - 2x_3 &= 0 \\ -2x_1 + x_2 - 3x_3 &= 2. \end{aligned}$$

These may be represented as an equation connecting two column vectors:

$$\begin{bmatrix} 2x_1 - x_2 + 4x_3 \\ x_1 + 3x_2 - 2x_3 \\ -2x_1 + x_2 - 3x_3 \end{bmatrix} = \begin{bmatrix} 1 \\ 0 \\ 2 \end{bmatrix}.$$

The idea of multiplication of a matrix with a vector is defined so that the left-hand vector is the result of multiplying the vector of unknowns by the matrix of coefficients, thus:

$$\begin{bmatrix} 2 & -1 & 4 \\ 1 & 3 & -2 \\ -2 & 1 & -3 \end{bmatrix} \begin{bmatrix} x_1 \\ x_2 \\ x_3 \end{bmatrix} = \begin{bmatrix} 2x_1 - x_2 + 4x_3 \\ x_1 + 3x_2 - 2x_3 \\ -2x_1 + x_2 - 3x_3 \end{bmatrix}.$$

In this way the original set of simultaneous equations may be written as a matrix equation:

$$\begin{bmatrix} 2 & -1 & 4 \\ 1 & 3 & -2 \\ -2 & 1 & -3 \end{bmatrix} \begin{bmatrix} x_1 \\ x_2 \\ x_3 \end{bmatrix} = \begin{bmatrix} 1 \\ 0 \\ 2 \end{bmatrix}.$$

3.6 Examples of simultaneous equations written as matrix equations.

$$\begin{aligned} 3x_1 - 2x_2 &= 1 \\ 4x_1 + x_2 &= -2 \end{aligned} \qquad \begin{bmatrix} 3 & -2 \\ 4 & 1 \end{bmatrix} \begin{bmatrix} x_1 \\ x_2 \end{bmatrix} = \begin{bmatrix} 1 \\ -2 \end{bmatrix}.$$

$$\begin{aligned} x_1 + x_2 + x_3 &= 6 \\ x_1 - x_2 - x_3 &= 0 \end{aligned} \qquad \begin{bmatrix} 1 & 1 & 1 \\ 1 & -1 & -1 \end{bmatrix} \begin{bmatrix} x_1 \\ x_2 \\ x_3 \end{bmatrix} = \begin{bmatrix} 6 \\ 0 \end{bmatrix}.$$

$$\begin{aligned} 3x_1 - 2x_2 &= 0 \\ x_1 + x_2 &= 5 \\ -x_1 + 2x_2 &= 4 \end{aligned} \qquad \begin{bmatrix} 3 & -2 \\ 1 & 1 \\ -1 & 2 \end{bmatrix} \begin{bmatrix} x_1 \\ x_2 \end{bmatrix} = \begin{bmatrix} 0 \\ 5 \\ 4 \end{bmatrix}.$$

3.7 Multiplication of a column vector by a matrix.

$$\begin{bmatrix} 1 & 2 \\ 3 & 4 \end{bmatrix} \begin{bmatrix} x_1 \\ x_2 \end{bmatrix} = \begin{bmatrix} x_1 + 2x_2 \\ 3x_1 + 4x_2 \end{bmatrix}.$$

$$\begin{bmatrix} 1 & 2 \\ 3 & 4 \end{bmatrix} \begin{bmatrix} 5 \\ 6 \end{bmatrix} = \begin{bmatrix} 5 + 12 \\ 15 + 24 \end{bmatrix} = \begin{bmatrix} 17 \\ 39 \end{bmatrix}.$$

$$\begin{bmatrix} 1 & 2 & 1 \\ -1 & -3 & 2 \end{bmatrix} \begin{bmatrix} x_1 \\ x_2 \\ x_3 \end{bmatrix} = \begin{bmatrix} x_1 + 2x_2 + x_3 \\ -x_1 - 3x_2 + 2x_3 \end{bmatrix}.$$

$$\begin{bmatrix} 1 & 1 & 1 \\ -1 & 2 & 1 \\ 3 & 1 & 3 \end{bmatrix} \begin{bmatrix} 2 \\ -1 \\ -2 \end{bmatrix} = \begin{bmatrix} 2-1-2 \\ -2-2-2 \\ 6-1-6 \end{bmatrix} = \begin{bmatrix} -1 \\ -6 \\ -1 \end{bmatrix}.$$

In Examples 3.3 we see what happens when we add a matrix to itself. Each entry is added to itself. In other words, each entry is multiplied by 2. This obviously extends to the case where we add a matrix to itself three times or four times or any number of times. It is convenient, therefore, to introduce the idea of multiplication of a matrix (or a vector) by a number. Notice that the definition applies for any real number, not just for integers. To multiply a matrix by a number, just multiply each entry by the number. In formal terms, if A is the $p \times q$ matrix whose (i, j)-entry is a_{ij} and if k is any number, then kA is the $p \times q$ matrix whose (i, j)-entry is ka_{ij}. See Examples 3.4.

Multiplication of a matrix with a vector or with another matrix is more complicated. Example 3.5 provides some motivation. The three left-hand sides are taken as a column vector, and this column vector is the result of multiplying the 3×3 matrix of coefficients with the 3×1 matrix (3-vector) of the unknowns. In general:

$$\begin{bmatrix} a_{11} & a_{12} & a_{13} \\ a_{21} & a_{22} & a_{23} \\ a_{31} & a_{32} & a_{33} \end{bmatrix} \begin{bmatrix} x_1 \\ x_2 \\ x_3 \end{bmatrix} = \begin{bmatrix} a_{11}x_1 + a_{12}x_2 + a_{13}x_3 \\ a_{21}x_1 + a_{22}x_2 + a_{23}x_3 \\ a_{31}x_1 + a_{32}x_2 + a_{33}x_3 \end{bmatrix}.$$

Note that the right-hand side is a column vector. Further illustrations are given in Examples 3.6. This idea can be applied to any set of simultaneous equations, no matter how many unknowns or how many equations. The left-hand side can be represented as a product of a matrix with a column vector. A set of p equations in q unknowns involves a $p \times q$ matrix multiplied to a q-vector.

Now let us abstract the idea. Can we multiply any matrix with any column vector? Not by the above process. To make that work, there must be as many columns in the matrix as there are entries in the column vector. A $p \times q$ matrix can be multiplied on the right by a column vector only if it has q entries. The result of the multiplication is then a column vector with p entries. We just reverse the above process. See Examples 3.7.

3.8 Evaluate the product

$$\begin{bmatrix} 1 & 2 & 3 \\ 2 & 3 & 4 \\ 4 & 5 & 6 \end{bmatrix} \begin{bmatrix} 1 & -1 \\ 3 & -2 \\ -1 & 1 \end{bmatrix}.$$

The product is a 3×2 matrix. The first column of the product is

$$\begin{bmatrix} 1 & 2 & 3 \\ 2 & 3 & 4 \\ 4 & 5 & 6 \end{bmatrix} \begin{bmatrix} 1 \\ 3 \\ -1 \end{bmatrix}, \quad \text{i.e.} \quad \begin{bmatrix} 1+6-3 \\ 2+9-4 \\ 4+15-6 \end{bmatrix}, \quad \text{i.e.} \quad \begin{bmatrix} 4 \\ 7 \\ 13 \end{bmatrix}.$$

The second column of the product is

$$\begin{bmatrix} 1 & 2 & 3 \\ 2 & 3 & 4 \\ 4 & 5 & 6 \end{bmatrix} \begin{bmatrix} -1 \\ -2 \\ 1 \end{bmatrix}, \quad \text{i.e.} \quad \begin{bmatrix} -1-4+3 \\ -2-6+4 \\ -4-10+6 \end{bmatrix}, \quad \text{i.e.} \quad \begin{bmatrix} -2 \\ -4 \\ -8 \end{bmatrix}.$$

Hence the product matrix is

$$\begin{bmatrix} 4 & -2 \\ 7 & -4 \\ 13 & -8 \end{bmatrix}.$$

3.9 Evaluation of matrix products.

(i) $$\begin{bmatrix} 1 & 2 \\ 3 & 4 \end{bmatrix} \begin{bmatrix} 1 & 0 & -1 \\ 0 & 1 & -1 \end{bmatrix} = \begin{bmatrix} 1+0 & 0+2 & -1-2 \\ 3+0 & 0+4 & -3-4 \end{bmatrix} = \begin{bmatrix} 1 & 2 & -3 \\ 3 & 4 & -7 \end{bmatrix}.$$

(ii) $$\begin{bmatrix} 1 & -1 \end{bmatrix} \begin{bmatrix} 2 & 1 \\ 1 & 1 \end{bmatrix} = \begin{bmatrix} 2-1 & 1+1 \end{bmatrix} = \begin{bmatrix} 1 & 2 \end{bmatrix}.$$

(iii) $$\begin{bmatrix} 1 & 0 & 1 \\ 0 & 1 & 1 \\ 1 & 1 & 0 \end{bmatrix} \begin{bmatrix} 0 & 0 & 1 \\ 0 & 1 & 0 \\ 1 & 0 & 0 \end{bmatrix} = \begin{bmatrix} 0+0+1 & 0+0+0 & 1+0+0 \\ 0+0+1 & 0+1+0 & 0+0+0 \\ 0+0+0 & 0+1+0 & 1+0+0 \end{bmatrix} = \begin{bmatrix} 1 & 0 & 1 \\ 1 & 1 & 0 \\ 0 & 1 & 1 \end{bmatrix}.$$

(iv) $$\begin{bmatrix} 1 & 0 & 1 & 1 \\ 0 & 2 & -1 & 3 \end{bmatrix} \begin{bmatrix} 0 & 0 \\ 1 & 2 \\ -1 & 1 \\ 3 & -2 \end{bmatrix} = \begin{bmatrix} 0+0-1+3 & 0+0+1-2 \\ 0+2+1+9 & 0+4-1-6 \end{bmatrix} = \begin{bmatrix} 2 & -1 \\ 12 & -3 \end{bmatrix}.$$

Next we take this further, and say what is meant by the product of two matrices. The process is illustrated by Example 3.8. The columns of the product matrix are calculated in turn by finding the products of the left-hand matrix with, separately, each of the columns of the right-hand matrix. Let A be a $p \times q$ matrix whose (i, j)-entry is a_{ij}, and let B be a $q \times r$ matrix whose (i, j)-entry is b_{ij}. Then the product AB is a $p \times r$ matrix whose (i, j)-entry is $\sum_{k=1}^{q} a_{ik}b_{kj}$, i.e. the sum of all the products of the entries in the ith row of A with the respective entries in the jth column of B.

Rule

A $p \times q$ matrix can be multiplied on the right only by a matrix with q rows. If A is a $p \times q$ matrix and B is a $q \times r$ matrix, then the product AB is a $p \times r$ matrix.

There is a useful mnemonic here. We can think of matrices as dominoes. A p, q domino can be laid next to a q, r domino, and the resulting 'free' numbers are p and r.

Examples 3.9 illustrate the procedures in calculating products. It is important to notice that given matrices can be multiplied only if they have appropriate sizes, and that it may be possible to multiply matrices in one order but not in the reverse order.

The most important case of matrix multiplication is multiplication of a matrix by a column vector, so before we move on to consider properties of the general multiplication, let us recap the application to simultaneous equations. A set of simultaneous equations containing p equations in q unknowns can always be represented as a matrix equation of the form

$$A\boldsymbol{x} = \boldsymbol{h},$$

where A is a $p \times q$ matrix, $\boldsymbol{x}$ is a q-vector whose entries are the unknowns, and $\boldsymbol{h}$ is the p-vector whose entries are the right-hand sides of the given equations.

Rules

(i) $A + B = B + A$
(ii) $(A + B) + C = A + (B + C)$
(iii) $k(A + B) = kA + kB$
(iv) $(kA)B = k(AB)$
(v) $(AB)C = A(BC)$
(vi) $A(B + C) = AB + AC$
(vii) $(A + B)C = AC + BC$,

where A, B and C are any matrices whose sizes permit the formation of these sums and products, and k is any real number.

3.10 Show that for any $p\times q$ matrix A and any $q\times r$ matrices B and C,
$$A(B+C)=AB+AC.$$

Let a_{ij} denote the (i,j)-entry in A, for $1\leqslant i\leqslant p$ and $1\leqslant j\leqslant q$, let b_{ij} denote the (i,j)-entry in B, for $1\leqslant i\leqslant q$ and $1\leqslant j\leqslant r$, and let c_{ij} denote the (i,j)-entry in C, for $1\leqslant i\leqslant q$ and $1\leqslant j\leqslant r$. The (k,j)-entry in $B+C$ is then $b_{kj}+c_{kj}$. By the definition, then, the (i,j)-entry in $A(B+C)$ is
$$\sum_{k=1}^{q} a_{ik}(b_{kj}+c_{kj}),$$
i.e.
$$\sum_{k=1}^{q} a_{ik}b_{kj}+\sum_{k=1}^{q} a_{ik}c_{kj},$$
which is just the sum of the (i,j)-entries in AB and in AC. Hence
$$A(B+C)=AB+AC.$$

3.11 The commutative law fails for matrix multiplication. Let
$$A=\begin{bmatrix}1 & 2\\ 3 & 4\end{bmatrix},$$
and let
$$B=\begin{bmatrix}1 & 1\\ 0 & 1\end{bmatrix}.$$
Certainly both products AB and BA exist. Their values are different, however, as we can verify by direct calculation.
$$AB=\begin{bmatrix}1 & 2\\ 3 & 4\end{bmatrix}\begin{bmatrix}1 & 1\\ 0 & 1\end{bmatrix}=\begin{bmatrix}1 & 3\\ 3 & 7\end{bmatrix},$$
and
$$BA=\begin{bmatrix}1 & 1\\ 0 & 1\end{bmatrix}\begin{bmatrix}1 & 2\\ 3 & 4\end{bmatrix}=\begin{bmatrix}4 & 6\\ 3 & 4\end{bmatrix}.$$

Rules (i), (ii), (iii) and (iv) are easy to verify. They reflect corresponding properties of numbers, since the operations involved correspond to simple operations on the entries of the matrices. Rules (v), (vi) and (vii), while being convenient and familiar, are by no means obviously true. Proofs of them are intricate, but require no advanced methods. To illustrate the ideas, the proof of (vi) is given as Example 3.10.

There is one algebraic rule which is conspicuously absent from the above list. Multiplication of matrices does not satisfy the commutative law. The products AB and BA, even if they can both be formed, in general are not the same. See Example 3.11. This can lead to difficulties unless we are careful, particularly when multiplying out bracketed expressions. Consider the following:

$$(A+B)(A+B)=AA+AB+BA+BB,$$

so

$$(A+B)^2=A^2+AB+BA+B^2,$$

and the result must be left in this form, different from the usual expression for the square of a sum.

Finally a word about notation. Matrices we denote by upper case letters $A, B, C, \ldots, X, Y, Z, \ldots$. Column vectors we denote by bold-face lower case letters $\boldsymbol{a}, \boldsymbol{b}, \boldsymbol{c}, \ldots, \boldsymbol{x}, \boldsymbol{y}, \boldsymbol{z}, \ldots$. Thankfully, this is one situation where there is a notation which is almost universal.

Summary

Procedures for adding and multiplying vectors and matrices are given, together with rules for when sums and products can be formed. The algebraic laws satisfied by these operations are listed. It is shown how to write a set of simultaneous linear equations as a matrix equation.

Exercises

1. In each case below, evaluate the matrices $A+B$, $A\boldsymbol{x}$, $B\boldsymbol{x}$, $3A$, $\frac{1}{2}B$, where A, B and $\boldsymbol{x}$ are as given.

(i) $A=\begin{bmatrix}1 & 2\\ 3 & 4\end{bmatrix}$, $B=\begin{bmatrix}1 & -1\\ -1 & 1\end{bmatrix}$, $\boldsymbol{x}=\begin{bmatrix}x_1\\ x_2\end{bmatrix}$.

(ii) $A=\begin{bmatrix}3 & 0\\ 1 & 1\end{bmatrix}$, $B=\begin{bmatrix}-2 & 1\\ 0 & 2\end{bmatrix}$, $\boldsymbol{x}=\begin{bmatrix}1\\ 4\end{bmatrix}$.

(iii) $A=\begin{bmatrix}1 & -1 & 2\\ 0 & 1 & 1\\ 2 & 3 & -3\end{bmatrix}$, $B=\begin{bmatrix}0 & 0 & -1\\ 1 & 2 & 2\\ -3 & 1 & 0\end{bmatrix}$, $\boldsymbol{x}=\begin{bmatrix}x_1\\ x_2\\ x_3\end{bmatrix}$.

(iv) $A=\begin{bmatrix}-2 & -1 & 1\\ 0 & 1 & 4\\ 6 & -2 & 1\end{bmatrix}$, $B=\begin{bmatrix}1 & 1 & 1\\ 1 & 1 & 1\\ 1 & 1 & 1\end{bmatrix}$, $\boldsymbol{x}=\begin{bmatrix}2\\ 1\\ -1\end{bmatrix}$.

2. Evaluate all the following products of a matrix with a vector.

(i) $\begin{bmatrix}1 & 2 & -2\\ 3 & -1 & -1\\ -2 & 2 & 0\end{bmatrix}\begin{bmatrix}1\\ 2\\ 3\end{bmatrix}$. (ii) $\begin{bmatrix}2 & 2\\ 1 & -2\\ 0 & 3\end{bmatrix}\begin{bmatrix}1\\ 1\end{bmatrix}$.

(iii) $\begin{bmatrix}1 & 1 & -1 & 3\\ 2 & -2 & 0 & 1\end{bmatrix}\begin{bmatrix}0\\ 1\\ 1\\ -1\end{bmatrix}$. (iv) $\begin{bmatrix}2 & 2\\ 1 & -2\\ 0 & 3\\ 3 & 0\\ 0 & 0\end{bmatrix}\begin{bmatrix}1\\ 1\end{bmatrix}$.

(v) $\begin{bmatrix}1 & 1 & 1 & 1 & 1\\ 1 & 1 & 1 & 1 & 1\\ 1 & 1 & 1 & 1 & 1\end{bmatrix}\begin{bmatrix}1\\ 2\\ 3\\ 4\\ 5\end{bmatrix}$. (vi) $\begin{bmatrix}1 & 0 & 0\\ 0 & 1 & 0\\ 0 & 0 & 1\end{bmatrix}\begin{bmatrix}-2\\ 1\\ 3\end{bmatrix}$.

3. Let

$$A=\begin{bmatrix}5 & 2 & 3\\ 2 & -3 & 4\end{bmatrix},\quad B=\begin{bmatrix}2 & -1 & 1 & 0\\ 0 & 2 & 2 & 2\\ 3 & 0 & -1 & 3\end{bmatrix},$$

$$C=\begin{bmatrix}1 & 0 & 2\\ 2 & -3 & 0\\ 0 & 0 & 3\\ 2 & 1 & 0\end{bmatrix},\quad D=\begin{bmatrix}2 & -1\\ 1 & 2\\ 3 & -2\end{bmatrix}.$$

Evaluate the products AB, AD, BC, CB and CD. Is there any other product of two of these matrices which exists? Evaluate any such. Evaluate the products $A(BC)$ and $(AB)C$.

4. Evaluate the following matrix products.

(i) $\begin{bmatrix}2 & 1\\ -1 & 1\\ 3 & 2\end{bmatrix}\begin{bmatrix}3 & 4\\ 5 & 2\end{bmatrix}$.

(ii) $$\begin{bmatrix} 1 & 0 & 1 & 0 & 1 \\ 2 & 3 & -1 & 1 & 0 \\ 2 & 1 & -2 & -1 & 1 \\ 1 & 4 & 3 & -3 & 1 \end{bmatrix} \times \begin{bmatrix} 2 & 2 & 1 \\ -1 & -1 & 2 \\ 0 & 1 & 0 \\ -3 & -2 & 2 \\ 1 & 1 & 1 \end{bmatrix}.$$

(iii) $$\begin{bmatrix} 0 & 1 \\ 1 & 0 \end{bmatrix}\begin{bmatrix} 1 & 2 \\ 3 & 4 \end{bmatrix}.$$

(iv) $$\begin{bmatrix} 2 & -1 & 0 \\ 1 & -1 & 4 \end{bmatrix}\begin{bmatrix} 0 & 1 & 1 \\ -2 & 3 & 3 \\ 1 & 2 & -1 \end{bmatrix}\begin{bmatrix} 1 \\ -1 \\ 1 \end{bmatrix}.$$

5. Obtain $A^3 - 2A^2 + A - I$, when

$$A = \begin{bmatrix} 1 & 1 & 2 \\ 1 & 1 & 1 \\ 2 & 1 & 1 \end{bmatrix}.$$

6. How must the sizes of matrices A and B be related in order for both of the products AB and BA to exist?

Examples

4.1 Properties of a zero matrix.

$$\begin{bmatrix} 0 & 0 \\ 0 & 0 \end{bmatrix} + \begin{bmatrix} a & b \\ c & d \end{bmatrix} = \begin{bmatrix} a & b \\ c & d \end{bmatrix},$$

$$\begin{bmatrix} 0 & 0 \\ 0 & 0 \end{bmatrix} \begin{bmatrix} a & b \\ c & d \end{bmatrix} = \begin{bmatrix} 0 & 0 \\ 0 & 0 \end{bmatrix},$$

$$\begin{bmatrix} a & b \\ c & d \end{bmatrix} \begin{bmatrix} 0 & 0 \\ 0 & 0 \end{bmatrix} = \begin{bmatrix} 0 & 0 \\ 0 & 0 \end{bmatrix},$$

$$\begin{bmatrix} 0 & 0 \\ 0 & 0 \end{bmatrix} \begin{bmatrix} a & b & c \\ d & e & f \end{bmatrix} = \begin{bmatrix} 0 & 0 & 0 \\ 0 & 0 & 0 \end{bmatrix}.$$

4.2 Properties of an identity matrix.

$$\begin{bmatrix} 1 & 0 \\ 0 & 1 \end{bmatrix} \begin{bmatrix} a & b \\ c & d \end{bmatrix} = \begin{bmatrix} a & b \\ c & d \end{bmatrix},$$

$$\begin{bmatrix} a & b \\ c & d \end{bmatrix} \begin{bmatrix} 1 & 0 \\ 0 & 1 \end{bmatrix} = \begin{bmatrix} a & b \\ c & d \end{bmatrix},$$

$$\begin{bmatrix} 1 & 0 & 0 \\ 0 & 1 & 0 \\ 0 & 0 & 1 \end{bmatrix} \begin{bmatrix} a & b & c \\ d & e & f \\ g & h & k \end{bmatrix} = \begin{bmatrix} a & b & c \\ d & e & f \\ g & h & k \end{bmatrix},$$

$$\begin{bmatrix} a & b & c \\ d & e & f \\ g & h & k \end{bmatrix} \begin{bmatrix} 1 & 0 & 0 \\ 0 & 1 & 0 \\ 0 & 0 & 1 \end{bmatrix} = \begin{bmatrix} a & b & c \\ d & e & f \\ g & h & k \end{bmatrix}.$$

4.3 Examples of diagonal matrices.

$$\begin{bmatrix} 3 & 0 \\ 0 & 2 \end{bmatrix}, \quad \begin{bmatrix} 3 & 0 \\ 0 & 0 \end{bmatrix}, \quad \begin{bmatrix} -1 & 0 \\ 0 & -1 \end{bmatrix},$$

$$\begin{bmatrix} 6 & 0 & 0 \\ 0 & -2 & 0 \\ 0 & 0 & 1 \end{bmatrix}, \quad \begin{bmatrix} 2 & 0 & 0 \\ 0 & 2 & 0 \\ 0 & 0 & 2 \end{bmatrix}, \quad \begin{bmatrix} 0 & 0 & 0 \\ 0 & 1 & 0 \\ 0 & 0 & 0 \end{bmatrix},$$

$$\begin{bmatrix} 1 & 0 & 0 & 0 \\ 0 & 2 & 0 & 0 \\ 0 & 0 & 3 & 0 \\ 0 & 0 & 0 & 4 \end{bmatrix}.$$

4.4 The following matrices are upper triangular.

$$\begin{bmatrix} 1 & 2 \\ 0 & 1 \end{bmatrix}, \quad \begin{bmatrix} 1 & 1 & 3 \\ 0 & -2 & 2 \\ 0 & 0 & -1 \end{bmatrix}, \quad \begin{bmatrix} 2 & 1 & 0 \\ 0 & 1 & 1 \\ 0 & 0 & 1 \end{bmatrix}, \quad \begin{bmatrix} 0 & 1 & 1 \\ 0 & 0 & 2 \\ 0 & 0 & 1 \end{bmatrix}.$$

The following matrices are lower triangular.

$$\begin{bmatrix} 1 & 0 \\ 2 & 1 \end{bmatrix}, \quad \begin{bmatrix} 1 & 0 & 0 \\ 2 & -2 & 0 \\ 3 & 1 & -1 \end{bmatrix}, \quad \begin{bmatrix} 2 & 0 & 0 \\ 1 & 1 & 0 \\ 0 & 1 & 1 \end{bmatrix}, \quad \begin{bmatrix} 0 & 0 & 0 \\ 1 & 0 & 0 \\ 1 & 2 & 1 \end{bmatrix}.$$

4

Special matrices

Example 4.1 shows the properties of a *zero* matrix. This form of special matrix does not raise any problems. A matrix which consists entirely of 0s (a zero matrix) behaves just as we would expect. We normally use 0 (zero) to denote a zero matrix, and **0** to denote a zero column vector. Of course we should bear in mind that there are many zero matrices having different sizes.

From matrices which act like zero we turn to matrices which act like 1. A square matrix which has 1s down the diagonal from top left to bottom right (this diagonal is called the main diagonal) and has 0s elsewhere is called an *identity matrix*. Example 4.2 shows the property which such matrices have, namely

$$AI = IA = A,$$

where I is an identity matrix and A is a square matrix of the same size. Notice that identity matrices are square and that there is one $p \times p$ identity matrix for each number p. We denote it by I_p, or just I if the size is not important.

There are other sorts of special matrix which are distinctive because of their algebraic properties or because of their appearance (or both). We describe some types here, although their significance will not be clear till later.

A *diagonal matrix* is a square matrix which has zero entries at all points off the main diagonal. One particular sort of diagonal matrix is an identity matrix. Other examples are given in Examples 4.3. Of course we do not insist that all the entries on the main diagonal are non-zero. We might even consider a zero matrix to be a diagonal matrix. The sum and product of two $p \times p$ diagonal matrices are $p \times p$ diagonal matrices.

The main diagonal divides a square matrix into two triangles. A square matrix which has zeros at all positions below the main diagonal is called an

4.5 Sums and products of triangular matrices.

(i) A sum of upper triangular matrices is upper triangular.

$$\begin{bmatrix} 1 & 1 & 3 \\ 0 & -2 & 2 \\ 0 & 0 & -1 \end{bmatrix} + \begin{bmatrix} 2 & 1 & 0 \\ 0 & 1 & 1 \\ 0 & 0 & 1 \end{bmatrix} = \begin{bmatrix} 3 & 2 & 3 \\ 0 & -1 & 3 \\ 0 & 0 & 0 \end{bmatrix}.$$

(ii) A product of upper triangular matrices is upper triangular.

$$\begin{bmatrix} 1 & 1 & 3 \\ 0 & -2 & 2 \\ 0 & 0 & -1 \end{bmatrix} \begin{bmatrix} 2 & 1 & 0 \\ 0 & 1 & 1 \\ 0 & 0 & 1 \end{bmatrix} = \begin{bmatrix} 2 & 2 & 4 \\ 0 & -2 & 0 \\ 0 & 0 & -1 \end{bmatrix}.$$

(iii) A product of lower triangular matrices is lower triangular.

$$\begin{bmatrix} 1 & 0 & 0 \\ 2 & -2 & 0 \\ 3 & 1 & 1 \end{bmatrix} \begin{bmatrix} 2 & 0 & 0 \\ 1 & 1 & 0 \\ 0 & 1 & 1 \end{bmatrix} = \begin{bmatrix} 2 & 0 & 0 \\ 2 & -2 & 0 \\ 7 & 2 & 1 \end{bmatrix}.$$

4.6 Examples of transposed matrices.

$$\begin{bmatrix} 1 & 2 \\ 3 & 4 \end{bmatrix}^{\mathrm{T}} = \begin{bmatrix} 1 & 3 \\ 2 & 4 \end{bmatrix}, \quad \begin{bmatrix} 1 & 2 & 3 \\ 4 & 5 & 6 \end{bmatrix}^{\mathrm{T}} = \begin{bmatrix} 1 & 4 \\ 2 & 5 \\ 3 & 6 \end{bmatrix},$$

$$[1 \quad 2 \quad 3]^{\mathrm{T}} = \begin{bmatrix} 1 \\ 2 \\ 3 \end{bmatrix}, \quad \text{and} \quad \begin{bmatrix} 1 \\ 2 \\ 3 \end{bmatrix}^{\mathrm{T}} = [1 \quad 2 \quad 3],$$

$$\begin{bmatrix} 3 & 1 & 2 \\ 1 & 1 & -4 \\ 2 & -4 & -1 \end{bmatrix}^{\mathrm{T}} = \begin{bmatrix} 3 & 1 & 2 \\ 1 & 1 & -4 \\ 2 & -4 & -1 \end{bmatrix},$$

so this matrix is symmetric.

4.7 A sum of symmetric matrices is symmetric. Let A and B be symmetric matrices with the same size. Then $A^{\mathrm{T}} = A$ and $B^{\mathrm{T}} = B$.

$$(A+B)^{\mathrm{T}} = A^{\mathrm{T}} + B^{\mathrm{T}} = A + B,$$

and so $A+B$ is symmetric. Here is a particular case:

$$A = \begin{bmatrix} 3 & 1 & 2 \\ 1 & 1 & -4 \\ 2 & -4 & -1 \end{bmatrix}, \quad B = \begin{bmatrix} 1 & 2 & -1 \\ 2 & 2 & 0 \\ -1 & 0 & 3 \end{bmatrix}.$$

Then A and B are both symmetric, and

$$A+B = \begin{bmatrix} 4 & 3 & 1 \\ 3 & 3 & -4 \\ 1 & -4 & 2 \end{bmatrix},$$

which is symmetric.

A product of two symmetric matrices is generally *not* symmetric. With A and B as above,

$$AB = \begin{bmatrix} 3 & 8 & 3 \\ 7 & 4 & -13 \\ -5 & -4 & -5 \end{bmatrix},$$

which is not symmetric.

upper triangular matrix. A square matrix which has zeros at all positions above the main diagonal is called a *lower triangular* matrix. A matrix of one or other of these kinds is called a triangular matrix. Such matrices have convenient properties which make them useful in some applications. But we can see now, as in Example 4.5, that sums and products of upper (lower) triangular matrices are upper (lower) triangular. Notice that when the GE process is applied to a square matrix the result is always an upper triangular matrix.

The main diagonal also plays a part in our next kind of special matrix. A square matrix is *symmetric* if reflection in the main diagonal leaves the matrix unchanged. In formal terms, if A is any matrix whose (i,j)-entry is a_{ij}, the *transpose* of A (denoted by A^{T}) is the matrix whose (i,j)-entry is a_{ji}, i.e. the matrix obtained by reflecting in the main diagonal. A is symmetric if $A^{\mathrm{T}} = A$. Notice that the rows of A^{T} are the columns of A, and vice versa. See Example 4.6. Such matrices figure prominently in more advanced work, but we can see now (Example 4.7) that sums of symmetric matrices are symmetric, but products in general are not. There are three important rules about transposes.

4.8 Let A and B be any $p \times p$ matrices. Then $(AB)^T = B^T A^T$. To see this, let the (i, j)-entries of A and B be denoted by a_{ij} and b_{ij} respectively. The (i, j)-entry in $(AB)^T$ is the (j, i)-entry in AB, which is

$$\sum_{k=1}^{p} a_{jk} b_{ki}.$$

The (i, j)-entry in $B^T A^T$ is

$$\sum_{k=1}^{p} b_{ik}^T a_{kj}^T,$$

where b_{ik}^T is the (i, k)-entry in B^T and a_{kj}^T is the (k, j)-entry in A^T. Now from the definition of the transpose, we have

$$b_{ik}^T = b_{ki} \quad \text{and} \quad a_{kj}^T = a_{jk}.$$

Hence the (i, j)-entry in $B^T A^T$ is

$$\sum_{k=1}^{p} b_{ki} a_{jk}, \quad \text{i.e.} \quad \sum_{k=1}^{p} a_{jk} b_{ki},$$

which is the same as the (i, j)-entry in $(AB)^T$. This proves the result.

4.9 Examples of skew-symmetric matrices.

(i) $$\begin{bmatrix} 0 & 2 \\ -2 & 0 \end{bmatrix}^T = \begin{bmatrix} 0 & -2 \\ 2 & 0 \end{bmatrix} = -\begin{bmatrix} 0 & 2 \\ -2 & 0 \end{bmatrix}.$$

(ii) $$\begin{bmatrix} 0 & 1 & 2 \\ -1 & 0 & -3 \\ -2 & 3 & 0 \end{bmatrix}^T = \begin{bmatrix} 0 & -1 & -2 \\ 1 & 0 & 3 \\ 2 & -3 & 0 \end{bmatrix} = -\begin{bmatrix} 0 & 1 & 2 \\ -1 & 0 & -3 \\ -2 & 3 & 0 \end{bmatrix}.$$

4.10 Examples of orthogonal matrices.

(i) Let $A = \begin{bmatrix} \frac{1}{\sqrt{2}} & \frac{1}{\sqrt{2}} \\ -\frac{1}{\sqrt{2}} & \frac{1}{\sqrt{2}} \end{bmatrix}$. Then $A^T = \begin{bmatrix} \frac{1}{\sqrt{2}} & -\frac{1}{\sqrt{2}} \\ \frac{1}{\sqrt{2}} & \frac{1}{\sqrt{2}} \end{bmatrix}$,

so

$$A^T A = \begin{bmatrix} \frac{1}{2}+\frac{1}{2} & \frac{1}{2}-\frac{1}{2} \\ \frac{1}{2}-\frac{1}{2} & \frac{1}{2}+\frac{1}{2} \end{bmatrix} = \begin{bmatrix} 1 & 0 \\ 0 & 1 \end{bmatrix},$$

and

$$AA^T = \begin{bmatrix} \frac{1}{2}+\frac{1}{2} & -\frac{1}{2}+\frac{1}{2} \\ -\frac{1}{2}+\frac{1}{2} & \frac{1}{2}+\frac{1}{2} \end{bmatrix} = \begin{bmatrix} 1 & 0 \\ 0 & 1 \end{bmatrix}.$$

Hence A is orthogonal.

(ii) Let $B = \begin{bmatrix} \frac{2}{3} & -\frac{1}{3} & \frac{2}{3} \\ \frac{2}{3} & \frac{2}{3} & -\frac{1}{3} \\ -\frac{1}{3} & \frac{2}{3} & \frac{2}{3} \end{bmatrix}$. Then $B^T = \begin{bmatrix} \frac{2}{3} & \frac{2}{3} & -\frac{1}{3} \\ -\frac{1}{3} & \frac{2}{3} & \frac{2}{3} \\ \frac{2}{3} & -\frac{1}{3} & \frac{2}{3} \end{bmatrix}$.

Then by direct evaluation we verify that $B^T B = I$ and $BB^T = I$.

Rules

(i) $(A^T)^T = A$.
(ii) $(A+B)^T = A^T + B^T$,
(iii) $(AB)^T = B^T A^T$.

The last of these is important because of the change of the order of the multiplication. Remember this! The first two are quite easy to justify. The third is rather intricate, though not essentially difficult. A proof is given in Example 4.8.

The transpose of a matrix A may be related to A in other ways. A *skew-symmetric* matrix is a matrix for which $A^T = -A$. See Example 4.9. An *orthogonal* matrix is a square matrix for which $A^T A = I$ and $AA^T = I$. See Example 4.10.

4.11 Examples of elementary matrices.

$$E_1 = \begin{bmatrix} 0 & 1 & 0 \\ 1 & 0 & 0 \\ 0 & 0 & 1 \end{bmatrix},$$

obtained by interchanging the first two rows of an identity matrix.

$$E_2 = \begin{bmatrix} 1 & 0 & 0 \\ 0 & 1 & 0 \\ 0 & 0 & 5 \end{bmatrix},$$

obtained by multiplying the third row of an identity matrix by 5.

$$E_3 = \begin{bmatrix} 1 & 3 & 0 \\ 0 & 1 & 0 \\ 0 & 0 & 1 \end{bmatrix},$$

obtained by adding three times the second row to the first row in an identity matrix.

4.12 Let

$$A = \begin{bmatrix} 1 & 2 & 3 \\ 4 & 5 & 6 \\ 7 & 8 & 9 \end{bmatrix}.$$

Check the effects of premultiplying A by E_1, E_2 and E_3 above.

$$E_1 A = \begin{bmatrix} 0 & 1 & 0 \\ 1 & 0 & 0 \\ 0 & 0 & 1 \end{bmatrix} \begin{bmatrix} 1 & 2 & 3 \\ 4 & 5 & 6 \\ 7 & 8 & 9 \end{bmatrix} = \begin{bmatrix} 4 & 5 & 6 \\ 1 & 2 & 3 \\ 7 & 8 & 9 \end{bmatrix}.$$

(The first two rows are interchanged.)

$$E_2 A = \begin{bmatrix} 1 & 0 & 0 \\ 0 & 1 & 0 \\ 0 & 0 & 5 \end{bmatrix} \begin{bmatrix} 1 & 2 & 3 \\ 4 & 5 & 6 \\ 7 & 8 & 9 \end{bmatrix} = \begin{bmatrix} 1 & 2 & 3 \\ 4 & 5 & 6 \\ 35 & 40 & 45 \end{bmatrix}.$$

(The third row is multiplied by 5.)

$$E_3 A = \begin{bmatrix} 1 & 3 & 0 \\ 0 & 1 & 0 \\ 0 & 0 & 1 \end{bmatrix} \begin{bmatrix} 1 & 2 & 3 \\ 4 & 5 & 6 \\ 7 & 8 & 9 \end{bmatrix} = \begin{bmatrix} 13 & 17 & 21 \\ 4 & 5 & 6 \\ 7 & 8 & 9 \end{bmatrix}.$$

(Three times the second row is added to the first row.)

Examples 4.11 illustrate the notion of *elementary* matrix. An elementary matrix is a square matrix which is obtained from an identity matrix by the application of a single elementary row operation (see Chapter 1). The significance of such matrices lies in the following. Let E be obtained from a $p \times p$ identity matrix by application of a single elementary row operation, and let A be any $p \times q$ matrix. Then the product matrix EA is the same as the matrix obtained from A by applying the same elementary row operation directly to it. Examples 4.12 illustrate this. Our knowledge of the GE process enables us to say: given any square matrix A, there exists a sequence $E_1, E_2, \ldots, E_r$ of elementary matrices such that the product $E_rE_{r-1} \ldots E_2E_1A$ is an upper triangular matrix. These elementary matrices correspond to the elementary row operations carried out in the course of the GE process. For an explicit case of this, see Example 4.13.

Another important property of elementary matrices arises from the preceding discussion. Let E be an elementary matrix, obtained from an identity matrix by application of a single elementary row operation. Certainly E can be converted back into the identity matrix by application of another elementary row operation. Let F be the elementary matrix corresponding (as above) to this elementary row operation. Then $FE = I$. The two row operations cancel each other out, and the two elementary matrices correspondingly combine to give the identity matrix. It is not hard to see that $EF = I$ here also. Such matrices are called inverses of each other. We shall discuss that idea at length later. Examples 4.14 show some elementary matrices and their inverses. The reader should check that their products are identity matrices. Also, from the definition of an orthogonal matrix it is apparent that an orthogonal matrix and its transpose are inverses of each other.

Summary

Various special kinds of matrices are described: zero matrices, identity, diagonal, triangular, symmetric, skew-symmetric, orthogonal and elementary matrices. Some algebraic properties of these are discussed. The transpose of a square matrix is defined, and rules for transposition of sums and products are given. The correspondence between elementary matrices and elementary row operations is pointed out.

4.13 Find a sequence $E_1, E_2, \ldots, E_r$ of elementary matrices such that the product $E_rE_{r-1}\ldots E_1A$ is an upper triangular matrix, where

$$A=\begin{bmatrix}0 & 1 & -3 & 2\\ 1 & 2 & 1 & 1\\ 1 & 1 & 4 & 2\end{bmatrix}.$$

We proceed with the standard GE process, noting the elementary matrix which corresponds to each row operation.

$$\begin{bmatrix}0 & 1 & -3 & 2\\ 1 & 2 & 1 & 1\\ 1 & 1 & 4 & 2\end{bmatrix}$$

$$\begin{bmatrix}1 & 2 & 1 & 1\\ 0 & 1 & -3 & 2\\ 1 & 1 & 4 & 2\end{bmatrix} \text{ interchange rows} \qquad E_1=\begin{bmatrix}0 & 1 & 0\\ 1 & 0 & 0\\ 0 & 0 & 1\end{bmatrix}$$

$$\begin{bmatrix}1 & 2 & 1 & 1\\ 0 & 1 & -3 & 2\\ 0 & -1 & 3 & 1\end{bmatrix} \ (3)-(1) \qquad E_2=\begin{bmatrix}1 & 0 & 0\\ 0 & 1 & 1\\ -1 & 0 & 1\end{bmatrix}$$

$$\begin{bmatrix}1 & 2 & 1 & 1\\ 0 & 1 & -3 & 2\\ 0 & 0 & 0 & 3\end{bmatrix} \ (3)+(2) \qquad E_3=\begin{bmatrix}1 & 0 & 0\\ 0 & 1 & 0\\ 0 & 1 & 1\end{bmatrix}$$

$$\begin{bmatrix}1 & 2 & 1 & 1\\ 0 & 1 & -3 & 2\\ 0 & 0 & 0 & 1\end{bmatrix} \ (3)\div 3 \qquad E_4=\begin{bmatrix}1 & 0 & 0\\ 0 & 1 & 0\\ 0 & 0 & \frac{1}{3}\end{bmatrix}.$$

Hence

$$E_4E_3E_2E_1A=\begin{bmatrix}1 & 2 & 1 & 1\\ 0 & 1 & -3 & 2\\ 0 & 0 & 0 & 1\end{bmatrix}.$$

4.14 Elementary matrices and their inverses.

$$\begin{bmatrix}0 & 1 & 0\\ 1 & 0 & 0\\ 0 & 0 & 1\end{bmatrix} \text{ has inverse } \begin{bmatrix}0 & 1 & 0\\ 1 & 0 & 0\\ 0 & 0 & 1\end{bmatrix}.$$

$$\begin{bmatrix}1 & 0 & 0\\ 0 & 1 & 0\\ 0 & 0 & 5\end{bmatrix} \text{ has inverse } \begin{bmatrix}1 & 0 & 0\\ 0 & 1 & 0\\ 0 & 0 & \frac{1}{5}\end{bmatrix}.$$

$$\begin{bmatrix}1 & 3 & 0\\ 0 & 1 & 0\\ 0 & 0 & 1\end{bmatrix} \text{ has inverse } \begin{bmatrix}1 & -3 & 0\\ 0 & 1 & 0\\ 0 & 0 & 0\end{bmatrix}.$$

The way to see these is to consider the effect of premultiplying by first one and then the other of each given pair. The second 'undoes' the effect of the first.

Exercises

1. Evaluate A^2, A^3, and A^4, where

$$A=\begin{bmatrix}1&1&1\\0&1&1\\0&0&1\end{bmatrix}.$$

Carry out the same calculations for the matrix

$$B=\begin{bmatrix}1&0&0\\1&1&0\\1&1&1\end{bmatrix}.$$

2. Let I be the 3×3 identity matrix. Show that $AI=A$ whenever A is a 2×3 matrix. Does this hold for any $p\times 3$ matrix A, irrespective of the value of p? Likewise, is it the case that $IB=B$ for every $3\times q$ matrix B?

3. In each case below, evaluate AB, where A and B are as given.

(i) $A=\begin{bmatrix}1&-1&2\\0&2&1\\0&0&-1\end{bmatrix}$, $B=\begin{bmatrix}0&1&1\\0&2&-2\\0&0&1\end{bmatrix}$.

(ii) $A=\begin{bmatrix}1&2&1\\0&1&-2\\0&0&1\end{bmatrix}$, $B=\begin{bmatrix}x_1\\x_2\\x_3\end{bmatrix}$.

(iii) $A=\begin{bmatrix}1&0&0\\2&-1&0\\-2&1&3\end{bmatrix}$, $B=\begin{bmatrix}-1&0&0\\1&2&0\\2&1&1\end{bmatrix}$.

(iv) $A=[1\ \ 2\ \ 3]$, $B=\begin{bmatrix}1&0&0\\1&1&0\\1&1&1\end{bmatrix}$.

4. Evaluate the product

$$[x_1\ \ x_2\ \ x_3]\begin{bmatrix}1&0&0\\-1&1&0\\3&-2&1\end{bmatrix}.$$

Hence find values of x_1, x_2 and x_3 for which the product is equal to $[-1\ \ 4\ \ 1]$.

5. Let A be any square matrix. Show that $A+A^{\mathrm{T}}$ is a symmetric matrix. Show also that the products $A^{\mathrm{T}}A$ and AA^{T} are symmetric matrices.

6. Which of the following matrices are symmetric, and which are skew-symmetric (and which are neither)?

$$\begin{bmatrix}1&2\\2&3\end{bmatrix},\ \begin{bmatrix}1&2\\-2&3\end{bmatrix},\ \begin{bmatrix}0&2\\-2&0\end{bmatrix},\ \begin{bmatrix}1&2\\3&1\end{bmatrix},\ \begin{bmatrix}1&0\\0&-1\end{bmatrix},$$

$$\begin{bmatrix}-1&0&1\\0&2&2\\2&1&-1\end{bmatrix},\ \begin{bmatrix}0&1&-2\\-1&0&3\\2&-3&0\end{bmatrix},$$

$$\begin{bmatrix}1&2&0\\-2&0&-1\\0&1&1\end{bmatrix},\ \begin{bmatrix}2&3&1\\3&0&-1\\1&-1&2\end{bmatrix},$$

$$\begin{bmatrix} 1 & 0 & 1 \\ 0 & 1 & 0 \\ 1 & 0 & 1 \end{bmatrix}, \quad \begin{bmatrix} 1 & 0 & -1 \\ 0 & 1 & 0 \\ 1 & 0 & 1 \end{bmatrix}, \quad \begin{bmatrix} 1 & 1 & 0 \\ 1 & 0 & -1 \\ 0 & -1 & -1 \end{bmatrix}.$$

7. Show that the following matrices are orthogonal.

$$\begin{bmatrix} \frac{1}{\sqrt{5}} & -\frac{2}{\sqrt{5}} \\ \frac{2}{\sqrt{5}} & \frac{1}{\sqrt{5}} \end{bmatrix}, \quad \begin{bmatrix} -\frac{3}{5} & \frac{4}{5} \\ \frac{4}{5} & \frac{3}{5} \end{bmatrix}, \quad \begin{bmatrix} 0 & \frac{2}{\sqrt{6}} & -\frac{1}{\sqrt{3}} \\ \frac{1}{\sqrt{2}} & \frac{1}{\sqrt{6}} & \frac{1}{\sqrt{3}} \\ -\frac{1}{\sqrt{2}} & \frac{1}{\sqrt{6}} & \frac{1}{\sqrt{3}} \end{bmatrix}.$$

8. Show that a product of two orthogonal matrices of the same size is an orthogonal matrix.

9. Describe in words the effect of premultiplying a 4×4 matrix by each of the elementary matrices below. Also in each case write down the elementary matrix which has the reverse effect.

(i) $\begin{bmatrix} 1 & 0 & 0 & 0 \\ 0 & 0 & 1 & 0 \\ 0 & 1 & 0 & 0 \\ 0 & 0 & 0 & 1 \end{bmatrix}$. (ii) $\begin{bmatrix} 1 & 0 & 2 & 0 \\ 0 & 1 & 0 & 0 \\ 0 & 0 & 1 & 0 \\ 0 & 0 & 0 & 1 \end{bmatrix}$.

(iii) $\begin{bmatrix} 1 & 0 & 0 & 0 \\ 0 & 1 & 0 & 0 \\ -3 & 0 & 1 & 0 \\ 0 & 0 & 0 & 1 \end{bmatrix}$. (iv) $\begin{bmatrix} 1 & 0 & 0 & 0 \\ 0 & -2 & 0 & 0 \\ 0 & 0 & 1 & 0 \\ 0 & 0 & 0 & 1 \end{bmatrix}$.

10. Apply the Gaussian elimination process to the matrix

$$A = \begin{bmatrix} 0 & 1 & 3 \\ 1 & 2 & -1 \\ 2 & 3 & 1 \end{bmatrix},$$

noting at each stage the elementary matrix corresponding to the row operation applied. Evaluate the product T of these elementary matrices and check your answer by evaluating the product TA (which should be the same as the result of the GE process).

11. Repeat Exercise 10, with the matrix

$$A = \begin{bmatrix} 1 & -1 & 2 & 1 \\ -1 & 3 & 0 & 1 \\ 2 & 1 & 1 & -1 \end{bmatrix}.$$

Examples

5.1 Show that the inverse of a matrix (if it exists) is unique.

Let $AB = BA = I$ (so that B satisfies the requirements for the inverse of A). Now suppose that $AX = XA = I$. Then

$$BAX = (BA)X = IX = X.$$

Also

$$BAX = B(AX) = BI = B.$$

Hence $X = B$. Consequently B is the only matrix with the properties of the inverse of A.

5.2 An example of a matrix which does not have an inverse is

$$\begin{bmatrix} 1 & -1 \\ -1 & 1 \end{bmatrix}.$$

There is no matrix B such that

$$\begin{bmatrix} 1 & -1 \\ -1 & 1 \end{bmatrix} B = I.$$

To see this, let

$$B = \begin{bmatrix} a & b \\ c & d \end{bmatrix}.$$

Then

$$\begin{bmatrix} 1 & -1 \\ -1 & 1 \end{bmatrix}\begin{bmatrix} a & b \\ c & d \end{bmatrix} = \begin{bmatrix} a-c & b-d \\ -a+c & -b+d \end{bmatrix}.$$

This cannot equal

$$\begin{bmatrix} 1 & 0 \\ 0 & 1 \end{bmatrix},$$

for if $a - c = 1$ then $-a + c = -1$, not 0.

5.3 Let A be a diagonal matrix, say $A = [a_{ij}]_{p \times p}$, with $a_{ij} = 0$ when $i \neq j$. Suppose also that for $1 \leqslant i \leqslant p$ we have $a_{ii} \neq 0$ (there are no 0s on the main diagonal). Then A is invertible.

To see this, we show that B is the inverse of A, where $B = [b_{ij}]_{p \times p}$ is the diagonal matrix with $b_{ii} = 1/a_{ii}$, for $1 \leqslant i \leqslant p$. Calculate the product AB. The (i, i)-entry in AB is

$$\sum_{k=1}^{p} a_{ik} b_{ki},$$

which is equal to $a_{ii}b_{ii}$, since for $k \neq i$ we have $a_{ik} = b_{ki} = 0$. By the choice of b_{ii}, then, $a_{ii}b_{ii} = 1$ for each i, and so $AB = I$. Similarly $BA = I$. We are assuming the result that a product of diagonal matrices is a diagonal matrix (see Example 4.4).

5

Matrix inverses

At the end of Chapter 4 we discovered matrices E and F with the property that $EF=I$ and $FE=I$, and we said that they were inverses of each other. Generally, if A is a square matrix and B is a matrix of the same size with $AB=I$ and $BA=I$, then B is said to be the *inverse* of A. The inverse of A is denoted by A^{-1}. Example 5.1 is a proof that the inverse of a matrix (if it exists at all) is unique. Example 5.2 gives a matrix which does not have an inverse. So we must take care: not every matrix has an inverse. A matrix which does have an inverse is said to be *invertible* (or non-singular). Note that an invertible matrix must be square. A square matrix which is not invertible is said to be *singular*.

Following our discussion in Chapter 4 we can say that every elementary matrix is invertible and every orthogonal matrix is invertible. Example 5.3 shows that every diagonal matrix with no zeros on the main diagonal is invertible. There is, however, a standard procedure for testing whether a given matrix is invertible, and, if it is, of finding its inverse. This process is described in this chapter. It is an extension of the GE process.

5.4   Let $A = \begin{bmatrix} 1 & 1 & 1 \\ 1 & 2 & 3 \\ 0 & 1 & 1 \end{bmatrix}$.

Find whether A is invertible and, if it is, find A^{-1}.

First carry out the standard GE process on A, at the same time performing the same operations on an identity matrix.

$$
\begin{array}{ccc|ccc}
1 & 1 & 1 & 1 & 0 & 0 \\
1 & 2 & 3 & 0 & 1 & 0 \\
0 & 1 & 1 & 0 & 0 & 1
\end{array}
$$

$$
\begin{array}{ccc|ccc}
1 & 1 & 1 & 1 & 0 & 0 \\
0 & 1 & 2 & -1 & 1 & 0 \\
0 & 1 & 1 & 0 & 0 & 1
\end{array}
\quad (2)-(1) \quad E_1 = \begin{bmatrix} 1 & 0 & 0 \\ -1 & 1 & 0 \\ 0 & 0 & 1 \end{bmatrix}
$$

$$
\begin{array}{ccc|ccc}
1 & 1 & 1 & 1 & 0 & 0 \\
0 & 1 & 2 & -1 & 1 & 0 \\
0 & 0 & -1 & 1 & -1 & 1
\end{array}
\quad (3)-(2) \quad E_2 = \begin{bmatrix} 1 & 0 & 0 \\ 0 & 1 & 0 \\ 0 & -1 & 1 \end{bmatrix}
$$

$$
\begin{array}{ccc|ccc}
1 & 1 & 1 & 1 & 0 & 0 \\
0 & 1 & 2 & -1 & 1 & 0 \\
0 & 0 & 1 & -1 & 1 & -1
\end{array}
\quad (3)\times -1 \quad E_3 = \begin{bmatrix} 1 & 0 & 0 \\ 0 & 1 & 0 \\ 0 & 0 & -1 \end{bmatrix}.
$$

This is where the standard process ends. The matrix A' referred to in the text is

$$
\begin{bmatrix} 1 & 1 & 1 \\ 0 & 1 & 2 \\ 0 & 0 & 1 \end{bmatrix}.
$$

The process of finding the inverse continues with further row operations, with the objective of transforming it into an identity matrix.

$$
\begin{array}{ccc|ccc}
1 & 1 & 0 & 2 & -1 & 1 \\
0 & 1 & 2 & -1 & 1 & 0 \\
0 & 0 & 1 & -1 & 1 & -1
\end{array}
\quad (1)-(3) \quad E_4 = \begin{bmatrix} 1 & 0 & -1 \\ 0 & 1 & 0 \\ 0 & 0 & 1 \end{bmatrix}
$$

$$
\begin{array}{ccc|ccc}
1 & 1 & 0 & 2 & -1 & 1 \\
0 & 1 & 0 & 1 & -1 & 2 \\
0 & 0 & 1 & -1 & 1 & -1
\end{array}
\quad (2)-2\times(3) \quad E_5 = \begin{bmatrix} 1 & 0 & 0 \\ 0 & 1 & -2 \\ 0 & 0 & 1 \end{bmatrix}
$$

$$
\begin{array}{ccc|ccc}
1 & 0 & 0 & 1 & 0 & -1 \\
0 & 1 & 0 & 1 & -1 & 2 \\
0 & 0 & 1 & -1 & 1 & -1
\end{array}
\quad (1)-(2) \quad E_6 = \begin{bmatrix} 1 & -1 & 0 \\ 0 & 1 & 0 \\ 0 & 0 & 1 \end{bmatrix}.
$$

The process has been successful, so A is invertible, and

$$
A^{-1} = \begin{bmatrix} 1 & 0 & -1 \\ 1 & -1 & 2 \\ -1 & 1 & -1 \end{bmatrix}.
$$

Now check (just this once) that this is equal to the product of the elementary matrices $E_6 E_5 E_4 E_3 E_2 E_1$. In normal applications of this process there is no need to keep a note of the elementary matrices used.

Example 5.4 illustrates the basis of the procedure. Starting with a square matrix A, the GE process leads to an upper triangular matrix, say A'. In the example, continue as follows. Subtract a multiple of the third row of A' from the second row in order to get 0 in the (2, 3)-place. Next, subtract a multiple of the third row from the first row in order to get 0 in the (1, 3)-place. Last, subtract a multiple of the second row from the first row in order to get 0 in the (1, 2)-place. By the GE process followed by this procedure we convert A into an identity matrix by elementary row operations. There exist, therefore, elementary matrices $E_1, E_2, \ldots, E_s$ such that

$$I = E_s E_{s-1} \ldots E_2 E_1 A.$$

Now if we set $B = E_s E_{s-1} \ldots E_2 E_1$, then we have $BA = I$. We shall show that $AB = I$ also. Let $F_1, F_2, \ldots, F_s$ be the inverses of $E_1, E_2, \ldots, E_s$ respectively. Then

$$\begin{aligned} F_1 F_2 \ldots F_s &= F_1 F_2 \ldots F_s I \\ &= F_1 F_2 \ldots F_s E_s E_{s-1} \ldots E_2 E_1 A \\ &= IA = A, \end{aligned}$$

since $F_s E_s = I$, $F_{s-1} E_{s-1} = I, \ldots, F_1 E_1 = I$. Consequently,

$$AB = F_1 F_2 \ldots F_s E_s E_{s-1} \ldots E_2 E_1 = I.$$

Hence B is the inverse of A. Our procedure for finding the inverse of A must therefore calculate for us the product $E_s E_{s-1} \ldots E_2 E_1$. This product can be written as $E_s E_{s-1} \ldots E_2 E_1 I$, and this gives the hint. We convert A to I by certain elementary row operations. The *same* row operations convert I into A^{-1} (if it exists). Explicitly,

$$\text{if} \quad I = E_s \ldots E_1 A \quad \text{then} \quad A^{-1} = E_s \ldots E_1 I.$$

5.5 Find the inverse of the matrix $\begin{bmatrix} 1 & 0 & 2 \\ 0 & 1 & 2 \\ 1 & 2 & 0 \end{bmatrix}$.

$$\begin{array}{rrr|rrrl}
1 & 0 & 2 & 1 & 0 & 0 & \\
0 & 1 & 2 & 0 & 1 & 0 & \\
1 & 2 & 0 & 0 & 0 & 1 & \\
\hline
1 & 0 & 2 & 1 & 0 & 0 & \\
0 & 1 & 2 & 0 & 1 & 0 & \\
0 & 2 & -2 & -1 & 0 & 1 & (3)-(1) \\
\hline
1 & 0 & 2 & 1 & 0 & 0 & \\
0 & 1 & 2 & 0 & 1 & 0 & \\
0 & 0 & -6 & -1 & -2 & 1 & (3)-2\times(2) \\
\hline
1 & 0 & 2 & 1 & 0 & 0 & \\
0 & 1 & 2 & 0 & 1 & 0 & \\
0 & 0 & 1 & \frac{1}{6} & \frac{1}{3} & -\frac{1}{6} & (3)\div -6 \\
\hline
\end{array}$$

(At this stage we can be sure that the given matrix *is* invertible, and that the process will succeed in finding the inverse.)

$$\begin{array}{rrr|rrrl}
1 & 0 & 0 & \frac{2}{3} & -\frac{2}{3} & \frac{1}{3} & (1)-2\times(3) \\
0 & 1 & 0 & -\frac{1}{3} & \frac{1}{3} & \frac{1}{3} & (2)-2\times(3) \\
0 & 0 & 1 & \frac{1}{6} & \frac{1}{3} & -\frac{1}{6} & \\
\hline
\end{array}$$

This is the end of the process, since the left-hand matrix is an identity matrix. We have shown that

$$\begin{bmatrix} 1 & 0 & 2 \\ 0 & 1 & 2 \\ 1 & 2 & 0 \end{bmatrix}^{-1} = \begin{bmatrix} \frac{2}{3} & -\frac{2}{3} & \frac{1}{3} \\ -\frac{1}{3} & \frac{1}{3} & \frac{1}{3} \\ \frac{1}{6} & \frac{1}{3} & -\frac{1}{6} \end{bmatrix}.$$

5.6 Find (if possible) the inverse of the matrix $\begin{bmatrix} 1 & 2 & 3 \\ 1 & 1 & 2 \\ 0 & 1 & 1 \end{bmatrix}$.

$$\begin{array}{rrr|rrrl}
1 & 2 & 3 & 1 & 0 & 0 & \\
1 & 1 & 2 & 0 & 1 & 0 & \\
0 & 1 & 1 & 0 & 0 & 1 & \\
\hline
1 & 2 & 3 & 1 & 0 & 0 & \\
0 & -1 & -1 & -1 & 1 & 0 & (2)-(1) \\
0 & 1 & 1 & 0 & 0 & 1 & \\
\hline
1 & 2 & 3 & 1 & 0 & 0 & \\
0 & 1 & 1 & 1 & -1 & 0 & (2)\div -1 \\
0 & 1 & 1 & 0 & 0 & 1 & \\
\hline
1 & 2 & 3 & 1 & 0 & 0 & \\
0 & 1 & 1 & 1 & -1 & 0 & \\
0 & 0 & 0 & -1 & 1 & 1 & (3)-(2) \\
\hline
\end{array}$$

The practical process for finding inverses is illustrated by Example 5.5. Apply elementary row operations to the given matrix A to convert it to I. At the same time, apply the same elementary row operations to I, thus converting it into A^{-1} (provided A^{-1} exists, as it does in this example). This shows how to find the inverse of a 3×3 matrix, but the method extends to any size of square matrix. Apply elementary row operations to obtain zeros below the main diagonal, as in the GE process, and, once this is complete, carry on with the procedure for obtaining zeros above the main diagonal as well. Remember that there is a simple way to check the answer when finding the inverse of a given matrix A. If your answer is B, calculate the product AB. It should be I. If it is not, then you have made a mistake.

What happens to our process for finding inverses if the original matrix A is not invertible? The method depended on obtaining, during the process, the matrix A' which had 1s on the main diagonal and 0s below it. As we saw in Chapter 2, this need not always be possible. It could happen that the last row (after the GE process) consists entirely of 0s. In such a case the process for finding the inverse breaks down at this point. There is no way to obtain 0s in the other places in the last column. Example 5.6 illustrates this. It is precisely in these cases that the original matrix is not invertible. We can see this quite easily. Suppose that the matrix A' has last row all 0s. There exist elementary matrices $E_1, E_2 \dots, E_r$ such that

$$A' = E_r E_{r-1} \dots E_2 E_1 A.$$

Now suppose (by way of contradiction) that A is invertible. Then $AA^{-1} = I$. Let $F_1, F_2, \dots, F_r$ be the inverses of $E_1, E_2, \dots, E_r$ respectively. Then

$$\begin{aligned} A'(A^{-1}F_1F_2 \dots F_r) &= (E_rE_{r-1} \dots E_2E_1A)A^{-1}F_1 \dots F_r \\ &= E_rE_{r-1} \dots E_2E_1IF_1 \dots F_r \\ &= I. \end{aligned}$$

Here the matrix A' is

$$\begin{bmatrix} 1 & 2 & 3 \\ 0 & 1 & 1 \\ 0 & 0 & 0 \end{bmatrix},$$

and the process for finding the inverse cannot be continued, because of the zero in the (3, 3)-place. The conclusion that we draw is that the given matrix is not invertible.

5.7 Let A be a $p \times p$ matrix whose pth (last) row consists entirely of 0s, and let X be any $p \times p$ matrix. Show the product AX has pth row consisting entirely of 0s.

Let $A = [a_{ij}]_{p \times p}$, with $a_{pj} = 0$ for $1 \leqslant j \leqslant p$. Let $X = [x_{ij}]_{p \times p}$. In AX the (p, j)-entry is $\sum_{k=1}^{p} a_{pk} x_{kj}$. But $a_{pk} = 0$ for all k, so this sum is zero. Hence the pth row of AX consists entirely of 0s.

5.8 A formula for the inverse of a 2×2 matrix.

$$\begin{bmatrix} a & b \\ c & d \end{bmatrix}^{-1} = \frac{1}{ad - bc} \begin{bmatrix} d & -b \\ -c & a \end{bmatrix},$$

provided that $ad - bc \neq 0$.

This is easily verified by multiplying out.

5.9 Show that if A and B are square matrices with $AB = I$, then A is invertible and $A^{-1} = B$.

Suppose that $AB = I$ and A is singular. Then, by the discussion in the text, there is an invertible matrix X such that XA has last row consisting of 0s. It follows that XAB has last row consisting of 0s (see Example 5.7). But $XAB = X$, since $AB = I$. But X cannot have its last row all 0s, because it is invertible (think about the process for finding the inverse). From this contradiction we may deduce that A is invertible. It remains to show that $A^{-1} = B$.

We have $AB = I$, and so

$$A^{-1}(AB) = A^{-1}I = A^{-1},$$

i.e. $$(A^{-1}A)B = A^{-1},$$

i.e. $$B = A^{-1}.$$

Notice that from $BA = I$ we can conclude by the same argument that B is invertible and $B^{-1} = A$. From this it follows that A is invertible and $A^{-1} = (B^{-1})^{-1} = B$.

Example 5.7 shows that such a product $A'X$, for any matrix X, has last row all 0s, and so $A'A^{-1}F_1 \ldots F_r$ has last row all 0s. But I does not. Hence the supposition that A is invertible is false.

Example 5.8 gives a formula for the inverse of a 2×2 matrix, if it exists. Example 5.10 is another calculation of an inverse.

Rule

A square matrix A is invertible if and only if the procedure given above reaches an intermediate stage with matrix A' having 1s on the main diagonal and 0s below it.

The definition of the matrix inverse required two conditions: B is the inverse of A if $AB = I$ *and* $BA = I$. It can be shown, however, that either one of these conditions is sufficient. Each condition implies the other. For a part proof of this, see Example 5.9. In practice it is very useful to use only one condition rather than two.

Next, a rule for inverses of products. Suppose that A and B are invertible $p \times p$ matrices. Must AB be invertible, and if so what is its inverse? Here is the trick:

$$\begin{aligned}(AB)(B^{-1}A^{-1}) &= A(BB^{-1})A^{-1} \quad \text{(rule (v) on page 27)}\\ &= AIA^{-1}\\ &= AA^{-1} = I,\end{aligned}$$

and

$$\begin{aligned}(B^{-1}A^{-1})(AB) &= B^{-1}(A^{-1}A)B\\ &= B^{-1}IB = I.\end{aligned}$$

Thus the matrix $B^{-1}A^{-1}$ has the required properties and, since inverses are unique if they exist, we have:

Rule

If A and B are both invertible $p \times p$ matrices, then so is AB, and $(AB)^{-1} = B^{-1}A^{-1}$.

(Take note of the change in the order of the multiplication.)

This rule extends to products of any number of matrices. The order reverses. Indeed, we have come across an example of this already. The elementary matrices $E_1, E_2, \ldots, E_r$ had inverses $F_1, F_2, \ldots, F_r$ respectively, and

$$(E_rE_{r-1} \ldots E_2E_1)(F_1F_2 \ldots F_{r-1}F_r) = I,$$

and

$$(F_1F_2 \ldots F_{r-1}F_r)(E_rE_{r-1} \ldots E_2E_1) = I,$$

so the inverse of $E_rE_{r-1} \ldots E_2E_1$ is $F_1F_2 \ldots F_{r-1}F_r$, and vice versa.

5.10 Find the inverse (if it exists) of the matrix

$$\begin{bmatrix} 2 & 1 & 1 \\ 1 & 0 & -1 \\ 1 & 3 & 2 \end{bmatrix}.$$

$$\begin{array}{ccc|ccc}
2 & 1 & 1 & 1 & 0 & 0 \\
1 & 0 & -1 & 0 & 1 & 0 \\
1 & 3 & 2 & 0 & 0 & 1 \\
\hline
1 & \frac{1}{2} & \frac{1}{2} & \frac{1}{2} & 0 & 0 \quad (1)\div 2 \\
1 & 0 & -1 & 0 & 1 & 0 \\
1 & 3 & 2 & 0 & 0 & 1 \\
\hline
1 & \frac{1}{2} & \frac{1}{2} & \frac{1}{2} & 0 & 0 \\
0 & -\frac{1}{2} & -\frac{3}{2} & -\frac{1}{2} & 1 & 0 \quad (2)-(1) \\
0 & \frac{5}{2} & \frac{3}{2} & -\frac{1}{2} & 0 & 1 \quad (3)-(1) \\
\hline
1 & \frac{1}{2} & \frac{1}{2} & \frac{1}{2} & 0 & 0 \\
0 & 1 & 3 & 1 & -2 & 0 \quad (2)\div -\frac{1}{2} \\
0 & \frac{5}{2} & \frac{3}{2} & -\frac{1}{2} & 0 & 1 \\
\hline
1 & \frac{1}{2} & \frac{1}{2} & \frac{1}{2} & 0 & 0 \\
0 & 1 & 3 & 1 & -2 & 0 \\
0 & 0 & -6 & -3 & 5 & 1 \quad (3)-\frac{5}{2}\times(2) \\
\hline
1 & \frac{1}{2} & \frac{1}{2} & \frac{1}{2} & 0 & 0 \\
0 & 1 & 3 & 1 & -2 & 0 \\
0 & 0 & 1 & \frac{1}{2} & -\frac{5}{6} & -\frac{1}{6} \quad (3)\div -6 \\
\hline
\end{array}$$

(And here we can say that our matrix is invertible.)

$$\begin{array}{ccc|ccc}
1 & \frac{1}{2} & 0 & \frac{1}{4} & \frac{5}{12} & \frac{1}{12} \quad (1)-\frac{1}{2}\times(3) \\
0 & 1 & 1 & -\frac{1}{2} & \frac{1}{2} & \frac{1}{2} \quad (2)-3\times(3) \\
0 & 0 & 1 & \frac{1}{2} & -\frac{5}{6} & -\frac{1}{6} \\
\hline
1 & 0 & 0 & \frac{1}{2} & \frac{1}{6} & -\frac{1}{6} \quad (1)-\frac{1}{2}\times(2) \\
0 & 1 & 0 & -\frac{1}{2} & \frac{1}{2} & \frac{1}{2} \\
0 & 0 & 1 & \frac{1}{2} & -\frac{5}{6} & -\frac{1}{6} \\
\hline
\end{array}$$

Hence

$$\begin{bmatrix} 2 & 1 & 1 \\ 1 & 0 & -1 \\ 1 & 3 & 2 \end{bmatrix}^{-1} = \begin{bmatrix} \frac{1}{2} & \frac{1}{6} & -\frac{1}{6} \\ -\frac{1}{2} & \frac{1}{2} & \frac{1}{2} \\ \frac{1}{2} & -\frac{5}{6} & -\frac{1}{6} \end{bmatrix}.$$

Summary

The definitions are given of invertible and singular matrices. A procedure is given for deciding whether a given matrix is invertible and, if it is, finding the inverse. The validity of the procedure is established. Also a rule is given for writing down inverses of products of invertible matrices.

Exercises

1. Find which of the following matrices are invertible and which are singular. Find the inverses of those which are invertible. Verify your answers.

(i) $\begin{bmatrix} 1 & 2 \\ 1 & 3 \end{bmatrix}$. (ii) $\begin{bmatrix} 1 & 0 \\ 3 & 1 \end{bmatrix}$, (iii) $\begin{bmatrix} -2 & -3 \\ 4 & 6 \end{bmatrix}$.

(iv) $\begin{bmatrix} 1 & 2 & 1 \\ 0 & 1 & 2 \\ 0 & 0 & 1 \end{bmatrix}$. (v) $\begin{bmatrix} 1 & -1 & 2 \\ -1 & 2 & -1 \\ 1 & -3 & 1 \end{bmatrix}$.

(vi) $\begin{bmatrix} 0 & 1 & 1 \\ 1 & 0 & 1 \\ 1 & 1 & 0 \end{bmatrix}$. (vii) $\begin{bmatrix} 1 & 2 & -1 \\ 2 & 2 & -4 \\ -1 & 0 & 3 \end{bmatrix}$.

(viii) $\begin{bmatrix} 1 & -2 & -1 \\ 0 & 3 & 4 \\ -3 & 1 & 1 \end{bmatrix}$. (ix) $\begin{bmatrix} 1 & -1 & 4 \\ 2 & 3 & 3 \\ 3 & 1 & 8 \end{bmatrix}$.

(x) $\begin{bmatrix} 2 & 3 & -2 & 3 \\ 1 & 0 & 2 & 1 \\ -1 & 1 & 4 & -2 \\ 3 & 0 & 0 & 4 \end{bmatrix}$. (xi) $\begin{bmatrix} 1 & 1 & 1 & 1 \\ -2 & 1 & 0 & 3 \\ 3 & 0 & -2 & 5 \\ 1 & -1 & -1 & -3 \end{bmatrix}$.

2. Find the inverse of the diagonal matrix

$$\begin{bmatrix} a & 0 & 0 \\ 0 & b & 0 \\ 0 & 0 & c \end{bmatrix},$$

where a, b and c are non-zero.

3. Let A, B and C be invertible matrices of the same size. Show that the inverse of the product ABC is $C^{-1}B^{-1}A^{-1}$.

4. Let $\boldsymbol{x}$ and $\boldsymbol{y}$ be p-vectors. Show that $\boldsymbol{x}\boldsymbol{y}^{\mathrm{T}}$ is a $p \times p$ matrix and is singular. Pick some vectors $\boldsymbol{x}$ and $\boldsymbol{y}$ at random and verify that $\boldsymbol{x}\boldsymbol{y}^{\mathrm{T}}$ is singular.

5. Show that, for any invertible matrix A,

$(A^{-1})^{\mathrm{T}}A^{\mathrm{T}} = I$ and $A^{\mathrm{T}}(A^{-1})^{\mathrm{T}} = I$.

Deduce that A^{T} is invertible and that its inverse is the transpose of A^{-1}. Deduce also that if A is symmetric then A^{-1} is also symmetric.

6. Let X and A be $p \times p$ matrices such that X is singular and A is invertible. Show that the products XA and AX are both singular. (Hint: suppose that an inverse matrix exists and derive a contradiction to the fact that X does not have an inverse.)

Examples

6.1 Illustrations of linear dependence.

(i) $\left(\begin{bmatrix}3\\-3\end{bmatrix},\begin{bmatrix}1\\-1\end{bmatrix}\right)$ is LD: $2\begin{bmatrix}3\\-3\end{bmatrix}-6\begin{bmatrix}1\\-1\end{bmatrix}=\begin{bmatrix}0\\0\end{bmatrix}$.

(ii) $\left(\begin{bmatrix}1\\2\end{bmatrix},\begin{bmatrix}-1\\3\end{bmatrix},\begin{bmatrix}2\\0\end{bmatrix}\right)$ is LD: $6\begin{bmatrix}1\\2\end{bmatrix}-4\begin{bmatrix}-1\\3\end{bmatrix}-5\begin{bmatrix}2\\0\end{bmatrix}=\begin{bmatrix}0\\0\end{bmatrix}$.

(iii) $\left(\begin{bmatrix}3\\2\\1\end{bmatrix},\begin{bmatrix}0\\1\\-1\end{bmatrix},\begin{bmatrix}5\\1\\4\end{bmatrix}\right)$ is LD: $-5\begin{bmatrix}3\\2\\1\end{bmatrix}+7\begin{bmatrix}0\\1\\-1\end{bmatrix}+3\begin{bmatrix}5\\1\\4\end{bmatrix}=\begin{bmatrix}0\\0\\0\end{bmatrix}$.

6.2 A list of two non-zero 2-vectors is LD if and only if each is a multiple of the other. Let $\begin{bmatrix}a_1\\b_1\end{bmatrix}$, $\begin{bmatrix}a_2\\b_2\end{bmatrix}$ be two non-zero 2-vectors.

First, suppose that they constitute a LD list. Then there exist numbers x_1 and x_2, not both zero, such that

$$x_1\begin{bmatrix}a_1\\b_1\end{bmatrix}+x_2\begin{bmatrix}a_2\\b_2\end{bmatrix}=\begin{bmatrix}0\\0\end{bmatrix}.$$

Without loss of generality, say $x_1\neq 0$. Then since $\begin{bmatrix}a_1\\b_1\end{bmatrix}\neq\mathbf{0}$, we must have $x_2\neq 0$ also. Consequently,

$$\begin{bmatrix}a_1\\b_1\end{bmatrix}=-(x_2/x_1)\begin{bmatrix}a_2\\b_2\end{bmatrix}\quad\text{and}\quad\begin{bmatrix}a_2\\b_2\end{bmatrix}=-(x_1/x_2)\begin{bmatrix}a_1\\b_1\end{bmatrix},$$

i.e. each is a multiple of the other.

Conversely, suppose that each is a multiple of the other, say

$$\begin{bmatrix}a_1\\b_1\end{bmatrix}=k\begin{bmatrix}a_2\\b_2\end{bmatrix}\quad\text{and}\quad\begin{bmatrix}a_2\\b_2\end{bmatrix}=(1/k)\begin{bmatrix}a_1\\b_1\end{bmatrix}.$$

Then

$$\begin{bmatrix}a_1\\b_1\end{bmatrix}-k\begin{bmatrix}a_2\\b_2\end{bmatrix}=\begin{bmatrix}0\\0\end{bmatrix},\ \text{and so}\ \left(\begin{bmatrix}a_1\\b_1\end{bmatrix},\begin{bmatrix}a_2\\b_2\end{bmatrix}\right)\ \text{is LD.}$$

6.3 Show that any list of three 2-vectors is LD. To show that

$$\left(\begin{bmatrix}a_1\\b_1\end{bmatrix},\begin{bmatrix}a_2\\b_2\end{bmatrix},\begin{bmatrix}a_3\\b_3\end{bmatrix}\right)\ \text{is LD,}$$

we seek numbers x_1, x_2 and x_3, not all zero, such that

$$x_1\begin{bmatrix}a_1\\b_1\end{bmatrix}+x_2\begin{bmatrix}a_2\\b_2\end{bmatrix}+x_3\begin{bmatrix}a_3\\b_3\end{bmatrix}=\begin{bmatrix}0\\0\end{bmatrix},$$

i.e.

$$\left.\begin{aligned}a_1x_1+a_2x_2+a_3x_3&=0\\b_1x_1+b_2x_2+b_3x_3&=0\end{aligned}\right\}.$$

In other words, we seek solutions other than $x_1=x_2=x_3=0$ to this set of simultaneous equations. These equations are consistent (because $x_1=x_2=x_3=0$ do satisfy them), so by the rules in Chapter 2 there are infinitely many solutions. Thus there do exist non-trivial solutions, and so the given list of vectors is LD.

6

Linear independence and rank

Examples 6.1 illustrate what is meant by linear dependence of a list of vectors. More formally: given a list of vectors $v_1, \ldots, v_k$ of the same size (i.e. all are $p \times 1$ matrices for the same p), a *linear combination* of these vectors is a sum of multiples of them, i.e. $x_1v_1 + x_2v_2 + \cdots + x_kv_k$, where $x_1, \ldots, x_k$ are any numbers. A list of vectors is said to be *linearly dependent* (abbreviated to LD) if there is some non-trivial linear combination of them which is equal to the zero vector. Of course, in a trivial way, we can always obtain the zero vector by taking all of the coefficients $x_1, \ldots, x_k$ to be 0. A *non-trivial* linear combination is one in which at least one of the coefficients is non-zero.

A list of vectors of the same size which is not linearly dependent is said to be *linearly independent* (abbreviated to LI).

Example 6.2 deals with the case of a list of two 2-vectors. A list of two non-zero 2-vectors is LD if and only if each is a multiple of the other. Example 6.3 deals with a list of three 2-vectors. Such a list is *always* LD. Why? Because a certain set of simultaneous equations *must* have a solution of a certain kind.

6.4 Find whether the list

$$\left(\begin{bmatrix}1\\2\\5\end{bmatrix}, \begin{bmatrix}2\\-2\\4\end{bmatrix}, \begin{bmatrix}1\\1\\4\end{bmatrix}\right)$$

is LI or LD.

Seek numbers x_1, x_2 and x_3, not all zero, such that

$$x_1\begin{bmatrix}1\\2\\5\end{bmatrix} + x_2\begin{bmatrix}2\\-2\\4\end{bmatrix} + x_3\begin{bmatrix}1\\1\\4\end{bmatrix} = \begin{bmatrix}0\\0\\0\end{bmatrix},$$

i.e.

$$\left.\begin{aligned} x_1+2x_2+\ x_3&=0\\ 2x_1-2x_2+\ x_3&=0\\ 5x_1+4x_2+4x_3&=0 \end{aligned}\right\}.$$

Apply the standard GE process (details omitted):

$$\begin{bmatrix}1 & 2 & 1 & 0\\ 2 & -2 & 1 & 0\\ 5 & 4 & 4 & 0\end{bmatrix} \to \begin{bmatrix}1 & 2 & 1 & 0\\ 0 & 1 & \frac{1}{6} & 0\\ 0 & 0 & 0 & 0\end{bmatrix}.$$

From this we conclude that the set of equations has infinitely many solutions, and so the given list of vectors is LD.

6.5 Find whether the list

$$\left(\begin{bmatrix}1\\2\\-1\end{bmatrix}, \begin{bmatrix}2\\2\\0\end{bmatrix}, \begin{bmatrix}1\\4\\3\end{bmatrix}\right)$$

is LI or LD.

Following the same procedure as in Example 6.4, we seek a non-trivial solution to

$$\begin{bmatrix}1 & 2 & 1\\ 2 & 2 & 4\\ -1 & 0 & 3\end{bmatrix}\begin{bmatrix}x_1\\x_2\\x_3\end{bmatrix} = \begin{bmatrix}0\\0\\0\end{bmatrix}$$

(here writing the three simultaneous equations as a matrix equation). Apply the standard GE process:

$$\begin{bmatrix}1 & 2 & 1 & 0\\ 2 & 2 & 4 & 0\\ -1 & 0 & 3 & 0\end{bmatrix} \to \begin{bmatrix}1 & 2 & 1 & 0\\ 0 & 1 & -1 & 0\\ 0 & 0 & 1 & 0\end{bmatrix}.$$

From this we conclude that there is a unique solution to the equation, namely $x_1 = x_2 = x_3 = 0$. Consequently there does not exist a non-trivial linear combination of the given vectors which is equal to the zero vector. The given list is therefore LI.

6.6 Find whether the list

$$\left(\begin{bmatrix}1\\-1\\5\end{bmatrix}, \begin{bmatrix}1\\2\\-1\end{bmatrix}, \begin{bmatrix}2\\1\\2\end{bmatrix}, \begin{bmatrix}1\\2\\7\end{bmatrix}\right)$$

is LI or LD.

For the moment, however, let us see precisely how linear dependence and simultaneous equations are connected. Consider three 3-vectors, as in Example 6.4. Let us seek to show that these vectors constitute a list which is LD (even though they may not). So we seek coefficients x_1, x_2, x_3 (not all zero) to make the linear combination equal to the zero vector. Now the vector equation which x_1, x_2 and x_3 must satisfy, if we separate out the corresponding entries on each side, becomes a set of three simultaneous equations in the unknowns x_1, x_2 and x_3. We can use our standard procedure (the GE process) to solve these equations. But there is a particular feature of these equations. The right-hand sides are all 0s, so, as we noted earlier, there certainly is one solution (at least), namely $x_1 = x_2 = x_3 = 0$. What we seek is another solution (any other solution), and from our earlier work we know that if there is to be another solution then there must be infinitely many solutions, since the only possibilities are: no solutions, a unique solution, and infinitely many solutions. What is more, we know what form the result of the GE process must take if there are to be infinitely many solutions. The last row must consist entirely of 0s. In Example 6.4 it does, so the given vectors are LD. In Example 6.5 it does not, so the given vectors are LI.

Because the right-hand sides of the equations are all 0s in calculations of this kind, we can neglect this column (or omit it, as we customarily shall). Referring to Chapter 2 we can see:

Rule

Let $v_1, \ldots, v_q$ be a list of p-vectors. To test whether this set is LD or LI, form a matrix A with the vectors $v_1, \ldots, v_q$ as columns (so that A is a $p \times q$ matrix) and carry out the standard GE process on A. If the resulting matrix has fewer than q non-zero rows then the given list of vectors is LD. Otherwise it is LI.

Example 6.6 shows what happens with a list of four 3-vectors. It will always turn out to be LD. The matrix after the GE process is bound to have fewer than four non-zero rows. This illustrates a general rule.

Here we reduce the working to the bare essentials. Apply the GE process to the matrix

$$\begin{bmatrix} 1 & 1 & 2 & 1 \\ -1 & 2 & 1 & 2 \\ 5 & -1 & 2 & 7 \end{bmatrix}.$$

We obtain (details omitted) the matrix

$$\begin{bmatrix} 1 & 1 & 2 & 1 \\ 0 & 1 & 1 & 1 \\ 0 & 0 & 1 & -4 \end{bmatrix}.$$

This matrix has fewer than four non-zero rows, so if we were proceeding as in the previous examples and seeking solutions to equations we would conclude that there were infinitely many solutions. Consequently the given list of vectors is LD.

6.7 Illustrations of calculations of ranks of matrices.

(i) $\begin{bmatrix} 1 & 2 & -1 \\ 2 & 2 & -4 \\ -1 & 0 & 3 \end{bmatrix}$ has rank 2.

GE process:

$$\begin{bmatrix} 1 & 2 & -1 \\ 2 & 2 & -4 \\ -1 & 0 & 3 \end{bmatrix} \to \begin{bmatrix} 1 & 2 & -1 \\ 0 & 1 & 1 \\ 0 & 0 & 0 \end{bmatrix},$$

a matrix with *two* non-zero rows.

(ii) $\begin{bmatrix} 1 & 2 \\ -2 & 1 \end{bmatrix}$ has rank 2.

GE process:

$$\begin{bmatrix} 1 & 2 \\ -2 & 1 \end{bmatrix} \to \begin{bmatrix} 1 & 2 \\ 0 & 1 \end{bmatrix} \quad \text{(two non-zero rows).}$$

(iii) $\begin{bmatrix} 1 & 1 & 1 & 1 \\ 2 & 2 & -1 & 1 \\ -1 & 7 & 5 & 2 \end{bmatrix}$ has rank 3.

GE process:

$$\begin{bmatrix} 1 & 1 & 1 & 2 \\ 2 & 2 & -1 & 1 \\ -1 & 7 & 5 & 2 \end{bmatrix} \to \begin{bmatrix} 1 & 1 & 1 & 2 \\ 0 & 1 & \frac{3}{4} & \frac{1}{2} \\ 0 & 0 & 1 & 1 \end{bmatrix} \quad \text{(three non-zero rows).}$$

(iv) $\begin{bmatrix} 1 & 1 & 1 \\ 1 & 2 & 3 \\ 0 & 1 & 1 \end{bmatrix}$ has rank 3.

GE process:

$$\begin{bmatrix} 1 & 1 & 1 \\ 1 & 2 & 3 \\ 0 & 1 & 1 \end{bmatrix} \to \begin{bmatrix} 1 & 1 & 1 \\ 0 & 1 & 2 \\ 0 & 0 & 1 \end{bmatrix} \quad \text{(three non-zero rows).}$$

(v) $\begin{bmatrix} 1 & 2 & -1 \\ 2 & 4 & -2 \\ -1 & -2 & 1 \end{bmatrix}$ has rank 1.

Rule
Any list of p-vectors which contains more than p distinct vectors is LD.

Following Chapter 5, we have another rule.

Rule
If a matrix is invertible then its columns form a LI list of vectors.

(Recall that a $p \times p$ matrix is invertible if and only if the standard GE process leads to a matrix with p non-zero rows.)

Another important idea is already implicit in the above. The *rank* of a matrix is the number of non-zero rows remaining after the standard GE process. Examples 6.7 show how ranks are calculated. It is obvious that the rank of a $p \times q$ matrix is necessarily less than or equal to p. It is also less than or equal to q. To see this, think about the shape of the matrix remaining after the GE process. It has 0s everywhere below the main diagonal, which starts at the top left. The largest possible number of non-zero rows occurs when the main diagonal itself contains no 0s, and in that case the first q rows are non-zero and the remaining rows are all 0s.

Consideration of rank is useful when stating criteria for equations to have particular sorts of solutions. We shall pursue this in Chapter 8.

In the meantime let us consider a first version of what we shall call the *Equivalence Theorem*, which brings together, through the GE process, all the ideas covered so far.

GE process:

$$\begin{bmatrix} 1 & 2 & -1 \\ 2 & 4 & -2 \\ -1 & -2 & 1 \end{bmatrix} \rightarrow \begin{bmatrix} 1 & 2 & -1 \\ 0 & 0 & 0 \\ 0 & 0 & 0 \end{bmatrix} \quad \text{(one non-zero row).}$$

6.8 Illustration of the Equivalence Theorem.

(i) $$A = \begin{bmatrix} 1 & 2 & 3 \\ 1 & 1 & 2 \\ 1 & -1 & 1 \end{bmatrix}.$$

The GE process leads to

$$\begin{bmatrix} 1 & 2 & 3 \\ 0 & 1 & 1 \\ 0 & 0 & 1 \end{bmatrix}.$$

From this we can tell that the rank of A is 3, that the columns of A form a LI list, and that the process for finding the inverse of A will succeed, so A is invertible.

(ii) $$A = \begin{bmatrix} 1 & 1 & 1 & 0 \\ 1 & 1 & 0 & 1 \\ 1 & 0 & 1 & 1 \\ 0 & 1 & 1 & 1 \end{bmatrix}.$$

The GE process leads to

$$\begin{bmatrix} 1 & 1 & 1 & 0 \\ 0 & 1 & 0 & -1 \\ 0 & 0 & 1 & -1 \\ 0 & 0 & 0 & 1 \end{bmatrix}.$$

Consequently the rank of A is 4, the columns of A form a LI list, and A is invertible.

(iii) $$A = \begin{bmatrix} 1 & 3 & -1 \\ -2 & 1 & -5 \\ 4 & 5 & 3 \end{bmatrix}.$$

The GE process leads to

$$\begin{bmatrix} 1 & 3 & -1 \\ 0 & 1 & -1 \\ 0 & 0 & 0 \end{bmatrix}.$$

Consequently the rank of A is 2 (not equal to 3), the columns of A form a list which is LD (not LI), and the process for finding the inverse of A will fail, so A is not invertible. Also the equation $Ax=0$ has infinitely many solutions, so all conditions of the Equivalence Theorem fail.

Theorem

Let A be a $p \times p$ matrix. The following are equivalent.

(i) A is invertible.

(ii) The rank of A is equal to p.

(iii) The columns of A form a LI list.

(iv) The set of simultaneous equations which can be written $A\mathbf{x}=\mathbf{0}$ has no solution other than $\mathbf{x}=\mathbf{0}$.

The justification for this theorem is that in each of these situations the GE process leads to a matrix with p non-zero rows. See Examples 6.8.

In Chapter 7 we shall introduce another equivalent condition, involving determinants.

Summary

The definitions are given of linear dependence and linear independence of lists of vectors. A method is given for testing linear dependence and independence. The idea of rank is introduced, and the equivalence of invertibility with conditions involving rank, linear independence and solutions to equations is demonstrated, via the GE process.

Exercises

1. Find numbers x and y such that

$$x\begin{bmatrix}3\\1\\1\end{bmatrix}+y\begin{bmatrix}2\\-1\\1\end{bmatrix}=\begin{bmatrix}1\\-3\\1\end{bmatrix}.$$

2. For each given list of vectors, find whether the third vector is a linear combination of the first two.

(i) $\left(\begin{bmatrix}1\\2\end{bmatrix},\begin{bmatrix}0\\1\end{bmatrix},\begin{bmatrix}-1\\1\end{bmatrix}\right)$ (ii) $\left(\begin{bmatrix}0\\0\end{bmatrix},\begin{bmatrix}1\\2\end{bmatrix},\begin{bmatrix}2\\3\end{bmatrix}\right)$.

(iii) $\left(\begin{bmatrix}1\\1\\2\end{bmatrix},\begin{bmatrix}0\\1\\0\end{bmatrix},\begin{bmatrix}-1\\5\\-2\end{bmatrix}\right)$.

(iv) $\left(\begin{bmatrix}2\\-1\\-3\end{bmatrix},\begin{bmatrix}1\\5\\-1\end{bmatrix},\begin{bmatrix}3\\0\\-5\end{bmatrix}\right)$.

3. Show that each of the following lists is linearly dependent, and in each case find a linear combination of the given vectors which is equal to the zero vector.

(i) $\left(\begin{bmatrix}3\\-1\end{bmatrix},\begin{bmatrix}6\\-2\end{bmatrix}\right)$. (ii) $\left(\begin{bmatrix}-1\\1\end{bmatrix},\begin{bmatrix}0\\2\end{bmatrix},\begin{bmatrix}3\\4\end{bmatrix}\right)$.

(iii) $\left(\begin{bmatrix}-2\\1\\1\end{bmatrix},\begin{bmatrix}1\\1\\4\end{bmatrix},\begin{bmatrix}3\\-1\\0\end{bmatrix}\right)$.

(iv) $\left(\begin{bmatrix}1\\5\\9\end{bmatrix},\begin{bmatrix}-3\\1\\5\end{bmatrix},\begin{bmatrix}2\\3\\4\end{bmatrix}\right)$.

(v) $\left(\begin{bmatrix}0\\2\\6\end{bmatrix},\begin{bmatrix}3\\1\\6\end{bmatrix},\begin{bmatrix}4\\-2\\-2\end{bmatrix}\right)$.

(vi) $\left(\begin{bmatrix}1\\-2\\3\\8\end{bmatrix},\begin{bmatrix}2\\1\\1\\1\end{bmatrix},\begin{bmatrix}5\\3\\2\\1\end{bmatrix}\right)$.

4. Find, in each case below, whether the given list of vectors is linearly dependent or independent.

(i) $\left(\begin{bmatrix}2\\3\end{bmatrix},\begin{bmatrix}1\\-1\end{bmatrix}\right)$. (ii) $\left(\begin{bmatrix}1\\2\end{bmatrix},\begin{bmatrix}0\\0\end{bmatrix}\right)$.

(iii) $\left(\begin{bmatrix}3\\-1\\2\end{bmatrix},\begin{bmatrix}0\\0\\0\end{bmatrix},\begin{bmatrix}1\\0\\0\end{bmatrix}\right)$.

(iv) $\left(\begin{bmatrix}1\\1\\1\end{bmatrix},\begin{bmatrix}-2\\1\\2\end{bmatrix},\begin{bmatrix}1\\3\\4\end{bmatrix}\right)$.

(v) $\left(\begin{bmatrix}0\\1\\-1\end{bmatrix}, \begin{bmatrix}3\\4\\1\end{bmatrix}, \begin{bmatrix}-2\\2\\0\end{bmatrix}\right)$.

(vi) $\left(\begin{bmatrix}1\\-2\\-1\end{bmatrix}, \begin{bmatrix}3\\-1\\7\end{bmatrix}, \begin{bmatrix}1\\1\\5\end{bmatrix}\right)$.

(vii) $\left(\begin{bmatrix}1\\1\\2\\1\end{bmatrix}, \begin{bmatrix}-1\\0\\1\\-1\end{bmatrix}, \begin{bmatrix}2\\1\\1\\-2\end{bmatrix}, \begin{bmatrix}0\\1\\0\\0\end{bmatrix}\right)$.

5. Calculate the rank of each of the matrices given below.

$$\begin{bmatrix}3 & 6\\-1 & -2\end{bmatrix}, \begin{bmatrix}-1 & 0 & 3\\1 & 2 & 4\end{bmatrix}, \begin{bmatrix}2 & 1\\3 & -1\end{bmatrix}, \begin{bmatrix}1 & 0\\2 & 0\end{bmatrix},$$

$$\begin{bmatrix}-2 & 1 & 3\\1 & 1 & -1\\1 & 4 & 0\end{bmatrix}, \begin{bmatrix}1 & -3 & 2\\5 & 1 & 3\\9 & 5 & 4\end{bmatrix}, \begin{bmatrix}1 & -2 & 1\\1 & 1 & 3\\1 & 2 & 4\end{bmatrix},$$

$$\begin{bmatrix}0 & 3 & -2\\1 & 4 & 2\\-1 & 1 & 0\end{bmatrix}, \begin{bmatrix}1 & 2\\1 & -1\\0 & 2\end{bmatrix}, \begin{bmatrix}1 & 2\\-1 & -2\\3 & 6\end{bmatrix},$$

$$\begin{bmatrix}1 & 0\\2 & 0\\3 & 0\end{bmatrix}, \begin{bmatrix}2 & 2 & 3\\1 & -1 & 1\end{bmatrix}, \begin{bmatrix}1 & 0 & -2\\-2 & 0 & 4\end{bmatrix},$$

$$\begin{bmatrix}1 & -1 & 1 & 2\\3 & 1 & -2 & 0\\1 & 3 & -4 & -4\end{bmatrix}, \begin{bmatrix}0 & 1 & -1 & 1\\2 & 0 & 1 & 1\\-1 & 1 & 2 & 3\end{bmatrix},$$

$$\begin{bmatrix}1 & -1 & 2 & 0\\1 & 0 & 1 & 1\\2 & 1 & 1 & 0\\1 & -1 & -2 & 0\end{bmatrix}, \begin{bmatrix}1 & -1 & 2 & 0 & 3\\1 & 0 & 1 & 1 & 4\\2 & 1 & 1 & 0 & 2\\1 & -1 & -2 & 0 & -2\end{bmatrix},$$

$$\begin{bmatrix}-1 & 2 & 1 & 1\\0 & 1 & 2 & -1\\-1 & 1 & -1 & 2\\-1 & 3 & 3 & 0\end{bmatrix}, \begin{bmatrix}2 & 1 & -1 & 2\\1 & 2 & 1 & 2\\3 & 1 & -1 & 0\\0 & 2 & 1 & 4\end{bmatrix}.$$

6. Let $\boldsymbol{x}$ and $\boldsymbol{y}$ be 3-vectors. Then $\boldsymbol{x}\boldsymbol{y}^{\mathrm{T}}$ is a 3×3 matrix which is singular (see Chapter 5, Exercise 4). What is the rank of $\boldsymbol{x}\boldsymbol{y}^{\mathrm{T}}$? Try out some particular examples to see what happens. Does this result hold for p-vectors, for every p?

Examples

7.1 Evaluation of 2×2 determinants.

(i) $\begin{vmatrix} 1 & 2 \\ 3 & 4 \end{vmatrix} = 4-6=-2.$

(ii) $\begin{vmatrix} 3 & 1 \\ -2 & 5 \end{vmatrix} = 15-(-2)=17.$

(iii) $\begin{vmatrix} 0 & 1 \\ 4 & 5 \end{vmatrix} = 0-4=-4.$

(iv) $\begin{vmatrix} 3 & 0 \\ 2 & 7 \end{vmatrix} = 21-0=21.$

(v) $\begin{vmatrix} 0 & 1 \\ 1 & 0 \end{vmatrix} = 0-1=-1.$

(vi) $\begin{vmatrix} 0 & 0 \\ 2 & 5 \end{vmatrix} = 0-0=0.$

7

Determinants

A 2×2 determinant is written

$$\begin{vmatrix} a & b \\ c & d \end{vmatrix}.$$

What this means is just the number $ad-bc$. Examples 7.1 show some simple 2×2 determinants and their values. Determinants are not the same thing as matrices. A determinant has a numerical (or algebraic) value. A matrix is an array. However, it makes sense, given any 2×2 matrix A, to talk of *the determinant of* A, written det A.

$$\text{If} \quad A = \begin{bmatrix} a & b \\ c & d \end{bmatrix} \quad \text{then} \quad \det A = \begin{vmatrix} a & b \\ c & d \end{vmatrix} = ad - bc.$$

The significance and usefulness of determinants will be more apparent when we deal with 3×3 and larger determinants, but the reader will find this expression $ad-bc$ occurring previously in Example 5.8. Also it may be instructive (as an exercise) to go through the details of the solution (for x and y) of the simultaneous equations

$$\left.\begin{aligned} ax+by&=h \\ cx+dy&=k \end{aligned}\right\}.$$

A 3×3 determinant is written

$$\begin{vmatrix} a_1 & a_2 & a_3 \\ b_1 & b_2 & b_3 \\ c_1 & c_2 & c_3 \end{vmatrix}.$$

What this means is

$$a_1b_2c_3 + a_2b_3c_1 + a_3b_1c_2 - a_1b_3c_2 - a_2b_1c_3 - a_3b_2c_1.$$

Again, this is a number, calculated from the numbers in the given array. It makes sense, therefore, to talk of the determinant of A where A is a 3×3 matrix, with the obvious meaning.

7.2 (i) Evaluate the determinant

$$\begin{vmatrix} 1 & 2 & 3 \\ 0 & 1 & 2 \\ 1 & 0 & 3 \end{vmatrix}.$$

Here let us use the first method, with an extended array.

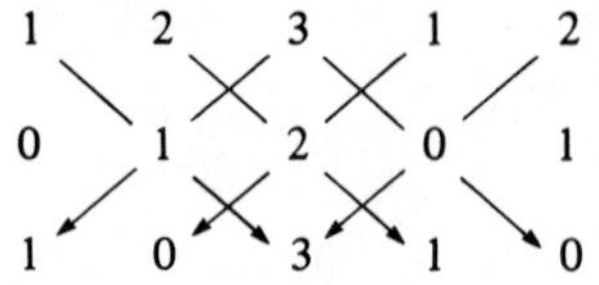

The value is $3+4+0-3-0-0$, i.e. 4.

(ii) Evaluate the determinant

$$\begin{vmatrix} 1 & -2 & -1 \\ 1 & -1 & -3 \\ 2 & -1 & 9 \end{vmatrix}.$$

Array:

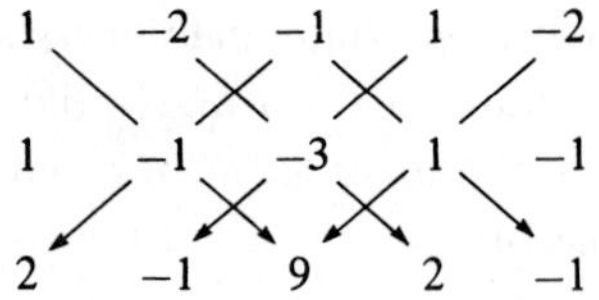

Value is: $-9+12+1-2-3-(-18)$, i.e. 17.

7.3 Evaluation of determinants using expansion by the first row.

(i) $$\begin{vmatrix} 1 & -2 & -1 \\ 1 & -1 & -3 \\ 2 & -1 & 9 \end{vmatrix} = 1\begin{vmatrix} -1 & -3 \\ -1 & 9 \end{vmatrix} - (-2)\begin{vmatrix} 1 & -3 \\ 2 & 9 \end{vmatrix} + (-1)\begin{vmatrix} 1 & -1 \\ 2 & -1 \end{vmatrix}$$

$$= (-12) + 2(15) - (1) = 17.$$

(ii) $$\begin{vmatrix} 1 & 2 & 3 \\ 0 & 1 & 2 \\ 1 & 0 & 3 \end{vmatrix} = 1\begin{vmatrix} 1 & 2 \\ 0 & 3 \end{vmatrix} - 2\begin{vmatrix} 0 & 2 \\ 1 & 3 \end{vmatrix} + 3\begin{vmatrix} 0 & 1 \\ 1 & 0 \end{vmatrix}$$

$$= 3 - 2(-2) + 3(-1) = 4.$$

(iii) $$\begin{vmatrix} 0 & 1 & 0 \\ 1 & 2 & 1 \\ 1 & 1 & 0 \end{vmatrix} = 0\begin{vmatrix} 2 & 1 \\ 1 & 0 \end{vmatrix} - 1\begin{vmatrix} 1 & 1 \\ 1 & 0 \end{vmatrix} + 0\begin{vmatrix} 1 & 2 \\ 1 & 1 \end{vmatrix}$$

$$= 0 - 1(-1) + 0 = 1.$$

(iv) $$\begin{vmatrix} 1 & 2 & 1 \\ 0 & 1 & 0 \\ 1 & 1 & 0 \end{vmatrix} = 1\begin{vmatrix} 1 & 0 \\ 1 & 0 \end{vmatrix} - 2\begin{vmatrix} 0 & 0 \\ 1 & 0 \end{vmatrix} + 1\begin{vmatrix} 0 & 1 \\ 1 & 1 \end{vmatrix}$$

$$= 0 - 2(0) + 1(-1) = -1.$$

How are we to cope with this imposing formula? There are several ways. Here is one way. Write down the given array, and then write the first and second columns again, on the right.

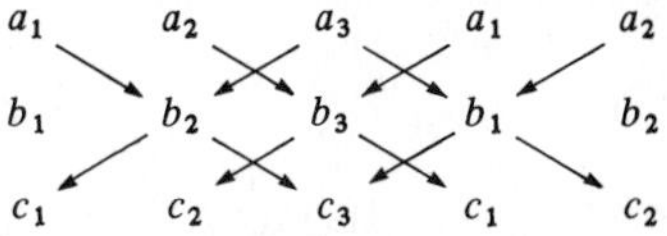

Add together the products of the numbers on the left-to-right arrows, and subtract the products of the numbers on the right-to-left arrows. A quick check with the definition above shows that we obtain the correct expression. Example 7.2 shows this method in operation.

The second method is called *expansion by a row or column*, and it is the key to the development of 4×4 and larger determinants so, although this method may seem more complicated, it is important and should be understood.

$$\begin{vmatrix} a_1 & a_2 & a_3 \\ b_1 & b_2 & b_3 \\ c_1 & c_2 & c_3 \end{vmatrix} = a_1\begin{vmatrix} b_2 & b_3 \\ c_2 & c_3 \end{vmatrix} - a_2\begin{vmatrix} b_1 & b_3 \\ c_1 & c_3 \end{vmatrix} + a_3\begin{vmatrix} b_1 & b_2 \\ c_1 & c_2 \end{vmatrix}.$$

Notice the form of the right-hand side. Each entry in the first row is multiplied with the determinant obtained by deleting the first row and the column containing that particular entry. To see that this gives the correct value for the determinant, we just have to multiply out the right-hand side as

$$a_1(b_2c_3-b_3c_2)-a_2(b_1c_3-b_3c_1)+a_3(b_1c_2-b_2c_1)$$

and compare with the original definition. When using this method, we must be careful to remember the negative sign which appears in front of the middle term. See Example 7.3.

There are similar expressions for a 3×3 determinant in which the coefficients are the entries in any of the rows or columns. We list these explicitly below.

$$-b_1\begin{vmatrix} a_2 & a_3 \\ c_2 & c_3 \end{vmatrix} + b_2\begin{vmatrix} a_1 & a_3 \\ c_1 & c_3 \end{vmatrix} - b_3\begin{vmatrix} a_1 & a_2 \\ c_1 & c_2 \end{vmatrix} \quad \text{(second row).}$$

$$c_1\begin{vmatrix} a_2 & a_3 \\ b_2 & b_3 \end{vmatrix} - c_2\begin{vmatrix} a_1 & a_3 \\ b_1 & b_3 \end{vmatrix} + c_3\begin{vmatrix} a_1 & a_2 \\ b_1 & b_2 \end{vmatrix} \quad \text{(third row).}$$

$$a_1\begin{vmatrix} b_2 & b_3 \\ c_2 & c_3 \end{vmatrix} - b_1\begin{vmatrix} a_2 & a_3 \\ c_2 & c_3 \end{vmatrix} + c_1\begin{vmatrix} a_2 & a_3 \\ b_2 & b_3 \end{vmatrix} \quad \text{(first column).}$$

$$-a_2\begin{vmatrix} b_1 & b_3 \\ c_1 & c_3 \end{vmatrix} + b_2\begin{vmatrix} a_1 & a_3 \\ c_1 & c_3 \end{vmatrix} - c_2\begin{vmatrix} a_1 & a_3 \\ b_1 & b_3 \end{vmatrix} \quad \text{(second column).}$$

7.4 Evaluation of determinants using expansion by a row or column.

(i)
$$\begin{vmatrix} 1 & 1 & 1 \\ 2 & 4 & -1 \\ 1 & -1 & 0 \end{vmatrix} = 1\begin{vmatrix} 1 & 1 \\ 4 & -1 \end{vmatrix} - (-1)\begin{vmatrix} 1 & 1 \\ 2 & -1 \end{vmatrix} + 0\begin{vmatrix} 1 & 1 \\ 2 & 4 \end{vmatrix}$$
$$= -5 - 3 = -8.$$

(Expansion by the third row.)

(ii)
$$\begin{vmatrix} 2 & 1 & 1 \\ 1 & 0 & -1 \\ 1 & 3 & 2 \end{vmatrix} = (-1)\begin{vmatrix} 1 & 1 \\ 3 & 2 \end{vmatrix} + 0\begin{vmatrix} 2 & 1 \\ 1 & 2 \end{vmatrix} - (-1)\begin{vmatrix} 2 & 1 \\ 1 & 3 \end{vmatrix}$$
$$= -(-1) + 5 = 6.$$

(Expansion by the second row.)

(iii)
$$\begin{vmatrix} 3 & 0 & 1 \\ -2 & -1 & 1 \\ 2 & 2 & -4 \end{vmatrix} = 0\begin{vmatrix} -2 & 1 \\ 2 & -4 \end{vmatrix} + (-1)\begin{vmatrix} 3 & 1 \\ 2 & -4 \end{vmatrix} - 2\begin{vmatrix} 3 & 1 \\ -2 & 1 \end{vmatrix}$$
$$= -(-14) - 2(5) = 4.$$

(Expansion by the second column.)

(iv)
$$\begin{vmatrix} 1 & 2 & 3 \\ 0 & 1 & 2 \\ 1 & 0 & 3 \end{vmatrix} = 1\begin{vmatrix} 1 & 2 \\ 0 & 3 \end{vmatrix} - 0\begin{vmatrix} 2 & 3 \\ 0 & 3 \end{vmatrix} + 1\begin{vmatrix} 2 & 3 \\ 1 & 2 \end{vmatrix}$$
$$= 3 + 1$$
$$= 4.$$

(Expansion by the first column.)

7.5 (i) Show that interchanging rows in a 3×3 determinant changes the sign of the value of the determinant.

We verify this by straight algebraic computation.

$$\begin{vmatrix} a_1 & a_2 & a_3 \\ b_1 & b_2 & b_3 \\ c_1 & c_2 & c_3 \end{vmatrix} = a_1\begin{vmatrix} b_2 & b_3 \\ c_2 & c_3 \end{vmatrix} - a_2\begin{vmatrix} b_1 & b_3 \\ c_1 & c_3 \end{vmatrix} + a_3\begin{vmatrix} b_1 & b_2 \\ c_1 & c_2 \end{vmatrix},$$

$$\begin{vmatrix} b_1 & b_2 & b_3 \\ a_1 & a_2 & a_3 \\ c_1 & c_2 & c_3 \end{vmatrix} = -a_1\begin{vmatrix} b_2 & b_3 \\ c_2 & c_3 \end{vmatrix} + a_2\begin{vmatrix} b_1 & b_3 \\ c_1 & c_3 \end{vmatrix} - a_3\begin{vmatrix} b_1 & b_2 \\ c_1 & c_2 \end{vmatrix}.$$

(The second determinant is evaluated by the second row.)

Similar calculations demonstrate the result for other possible single interchanges of rows.

(ii) Show that multiplying one row of a 3×3 determinant by a number k has the effect of multiplying the value of the determinant by k.

Again, straight computation gives this result.

$$\begin{vmatrix} a_1 & a_2 & a_3 \\ kb_1 & kb_2 & kb_3 \\ c_1 & c_2 & c_3 \end{vmatrix} = -kb_1\begin{vmatrix} a_2 & a_3 \\ c_2 & c_3 \end{vmatrix} + kb_2\begin{vmatrix} a_1 & a_3 \\ c_1 & c_3 \end{vmatrix} - kb_3\begin{vmatrix} a_1 & a_2 \\ c_1 & c_2 \end{vmatrix}$$
$$= k\left(-b_1\begin{vmatrix} a_2 & a_3 \\ c_2 & c_3 \end{vmatrix} + b_2\begin{vmatrix} a_1 & a_3 \\ c_1 & c_3 \end{vmatrix} - b_3\begin{vmatrix} a_1 & a_2 \\ c_1 & c_2 \end{vmatrix}\right)$$

$$a_3\begin{vmatrix} b_1 & b_2 \\ c_1 & c_2 \end{vmatrix} - b_3\begin{vmatrix} a_1 & a_2 \\ c_1 & c_2 \end{vmatrix} + c_3\begin{vmatrix} a_1 & a_2 \\ b_1 & b_2 \end{vmatrix} \quad \text{(third column).}$$

Again, justification of these is by multiplying out and comparing with the definition. It is important to notice the pattern of signs. Each entry always appears (as a coefficient) associated with a positive or negative sign, the same in each of the above expressions. This pattern is easy to remember from the array

$$\begin{bmatrix} + & - & + \\ - & + & - \\ + & - & + \end{bmatrix}.$$

Examples 7.4 give some further illustrations of evaluation of 3×3 determinants. Expansion by certain rows or columns can make the calculation easier in particular cases, especially when some of the entries are zeros.

Now let us try to connect these ideas with previous ideas. Recall the three kinds of elementary row operation given in Chapter 1, which form the basis of the GE process. How are determinants affected by these operations? We find the answers for 3×3 determinants.

Rule

(i) Interchanging two rows in a determinant changes the sign of the determinant.

(ii) Multiplying one row of a determinant by a number k has the effect of multiplying the determinant by k.

(iii) Adding a multiple of one row in a determinant to another row does not change the value of the determinant.

Proofs: (i) and (ii) are quite straightforward. See Examples 7.5. The third requires a little more discussion. Consider a particular case, as follows.

$$\begin{vmatrix} a_1 & a_2 & a_3 \\ b_1+kc_1 & b_2+kc_2 & b_3+kc_3 \\ c_1 & c_2 & c_3 \end{vmatrix}$$

$$= -(b_1+kc_1)\begin{vmatrix} a_2 & a_3 \\ c_2 & c_3 \end{vmatrix} + (b_2+kc_2)\begin{vmatrix} a_1 & a_3 \\ c_1 & c_3 \end{vmatrix}$$

$$-(b_3+kc_3)\begin{vmatrix} a_1 & a_2 \\ c_1 & c_2 \end{vmatrix}$$

$$=k\begin{vmatrix} a_1 & a_2 & a_3 \\ b_1 & b_2 & b_3 \\ c_1 & c_2 & c_3 \end{vmatrix}.$$

Similar calculations demonstrate the result when a multiplier is applied to another row.

7.6 Show that a 3×3 determinant in which two rows are identical has value zero.

Expansion by the other row gives the result:

$$\begin{vmatrix} a_1 & a_2 & a_3 \\ a_1 & a_2 & a_3 \\ c_1 & c_2 & c_3 \end{vmatrix} = c_1\begin{vmatrix} a_2 & a_3 \\ a_2 & a_3 \end{vmatrix} - c_2\begin{vmatrix} a_1 & a_3 \\ a_1 & a_3 \end{vmatrix} + c_3\begin{vmatrix} a_1 & a_2 \\ a_1 & a_2 \end{vmatrix}$$

$$= c_1(0) - c_2(0) + c_3(0)$$

$$= 0.$$

$$\begin{vmatrix} a_1 & a_2 & a_3 \\ b_1 & b_2 & b_3 \\ a_1 & a_2 & a_3 \end{vmatrix} = -b_1\begin{vmatrix} a_2 & a_3 \\ a_2 & a_3 \end{vmatrix} + b_2\begin{vmatrix} a_1 & a_3 \\ a_1 & a_3 \end{vmatrix} - b_3\begin{vmatrix} a_1 & a_2 \\ a_1 & a_2 \end{vmatrix}$$

$$= -b_1(0) + b_2(0) - b_3(0)$$

$$= 0.$$

(And similarly when the second and third rows are identical.)

7.7 Evaluate the 4×4 determinant

$$\begin{vmatrix} 2 & 0 & 1 & -1 \\ 1 & 1 & -1 & 0 \\ 0 & 3 & 1 & 3 \\ -2 & 1 & -1 & 1 \end{vmatrix}.$$

Expand by the first row, obtaining

$$2\begin{vmatrix} 1 & -1 & 0 \\ 3 & 1 & 3 \\ 1 & -1 & 1 \end{vmatrix} - 0\begin{vmatrix} 1 & -1 & 0 \\ 0 & 1 & 3 \\ -2 & -1 & 1 \end{vmatrix}$$

$$+1\begin{vmatrix} 1 & 1 & 0 \\ 0 & 3 & 3 \\ -2 & 1 & 1 \end{vmatrix} - (-1)\begin{vmatrix} 1 & 1 & -1 \\ 0 & 3 & 1 \\ -2 & 1 & -1 \end{vmatrix}$$

$$=2\left(\begin{vmatrix} 1 & 3 \\ -1 & 1 \end{vmatrix} + \begin{vmatrix} 3 & 3 \\ 1 & 1 \end{vmatrix}\right) + \left(\begin{vmatrix} 3 & 3 \\ 1 & 1 \end{vmatrix} - \begin{vmatrix} 0 & 3 \\ -2 & 1 \end{vmatrix}\right)$$

$$+\left(3\begin{vmatrix} 1 & -1 \\ -2 & -1 \end{vmatrix} - \begin{vmatrix} 1 & 1 \\ -2 & 1 \end{vmatrix}\right)$$

(the determinants being evaluated by the first row, the first row and the second row respectively)

$$= 2(4+0) + (0-6) + (-9-3)$$

$$= -10.$$

$$= -b_1\begin{vmatrix} a_2 & a_3 \\ c_2 & c_3 \end{vmatrix} + b_2\begin{vmatrix} a_1 & a_3 \\ c_1 & c_3 \end{vmatrix} - b_3\begin{vmatrix} a_1 & a_2 \\ c_1 & c_2 \end{vmatrix}$$

$$+k\left(-c_1\begin{vmatrix} a_2 & a_3 \\ c_2 & c_3 \end{vmatrix} + c_2\begin{vmatrix} a_1 & a_3 \\ c_1 & c_3 \end{vmatrix} - c_3\begin{vmatrix} a_1 & a_2 \\ c_1 & c_2 \end{vmatrix}\right)$$

$$= \begin{vmatrix} a_1 & a_2 & a_3 \\ b_1 & b_2 & b_3 \\ c_1 & c_2 & c_3 \end{vmatrix} + k(-c_1(a_2c_3 - a_3c_2) + c_2(a_1c_3 - a_3c_1) - c_3(a_1c_2 - a_2c_1))$$

$$= \begin{vmatrix} a_1 & a_2 & a_3 \\ b_1 & b_2 & b_3 \\ c_1 & c_2 & c_3 \end{vmatrix},$$

since the expression in brackets is identically zero (work it out!). This process works in exactly the same way for the other possible cases. Incidentally, we have come across another result of interest in the course of the above.

Rule

A determinant in which two rows are the same has value zero.

Proof: Notice that in the previous proof the expression in brackets is in fact the expansion by the second row of the determinant

$$\begin{vmatrix} a_1 & a_2 & a_3 \\ c_1 & c_2 & c_3 \\ c_1 & c_2 & c_3 \end{vmatrix},$$

and this is seen above to equal zero. Other possible cases work similarly. See Examples 7.6.

We shall not pursue the detailed discussion of larger determinants. A 4×4 determinant is defined most easily by means of expansion by the first row:

$$\begin{vmatrix} a_1 & a_2 & a_3 & a_4 \\ b_1 & b_2 & b_3 & b_4 \\ c_1 & c_2 & c_3 & c_4 \\ d_1 & d_2 & d_3 & d_4 \end{vmatrix} = a_1\begin{vmatrix} b_2 & b_3 & b_4 \\ c_2 & c_3 & c_4 \\ d_2 & d_3 & d_4 \end{vmatrix} - a_2\begin{vmatrix} b_1 & b_3 & b_4 \\ c_1 & c_3 & c_4 \\ d_1 & d_3 & d_4 \end{vmatrix}$$

$$+a_3\begin{vmatrix} b_1 & b_2 & b_4 \\ c_1 & c_2 & c_4 \\ d_1 & d_2 & d_4 \end{vmatrix} - a_4\begin{vmatrix} b_1 & b_2 & b_3 \\ c_1 & c_2 & c_3 \\ d_1 & d_2 & d_3 \end{vmatrix}.$$

See Example 7.7. Expansions by the other rows and by the columns produce the same value, provided we remember the alternating pattern of signs

7.8 Evaluation of determinants of triangular matrices.

(i) $\begin{vmatrix} a & b & c \\ 0 & d & e \\ 0 & 0 & f \end{vmatrix} = a\begin{vmatrix} d & e \\ 0 & f \end{vmatrix} = adf.$ (Expanding by the first column.)

(ii) $\begin{vmatrix} a_1 & a_2 & a_3 & a_4 \\ 0 & b_2 & b_3 & b_4 \\ 0 & 0 & c_3 & c_4 \\ 0 & 0 & 0 & d_4 \end{vmatrix} = a_1\begin{vmatrix} b_2 & b_3 & b_4 \\ 0 & c_3 & c_4 \\ 0 & 0 & d_4 \end{vmatrix}$ (by first column)

$= a_1(b_2c_3d_4)$ (by part (i))

$= a_1b_2c_3d_4.$

(iii) Larger determinants yield corresponding results. The determinant of *any* triangular matrix is equal to the product of the entries on the main diagonal. You should be able to see why this happens. A proper proof would use the Principle of Mathematical Induction. (We have dealt here with upper triangular matrices. Similar arguments apply in the case of lower triangular matrices.)

7.9 Evaluation of determinants using the GE process.

(i) $\begin{vmatrix} 1 & 3 & -1 \\ 2 & 0 & 1 \\ 1 & 1 & 4 \end{vmatrix}.$

Proceed with the GE process, noting the effect on the determinant of each row operation performed.

1	3	−1		Leaves the determinant unchanged.
0	−6	3	(2) − 2 × (1)	
0	−2	5	(3) − (1)	
1	3	−1		Divides the determinant by −6.
0	1	$-\frac{1}{2}$	(2) ÷ −6	
0	−2	5		
1	3	−1		Leaves the determinant unchanged.
0	1	$-\frac{1}{2}$		
0	0	4	(3) + 2 × (2)	

This last matrix is upper triangular, and has determinant equal to 4. Hence the original determinant has value $4 \times (-6)$, i.e. -24.

(ii) $\begin{vmatrix} 0 & 1 & 1 & 1 \\ 1 & 0 & 1 & 1 \\ 1 & 1 & 0 & 1 \\ 1 & 1 & 1 & 0 \end{vmatrix}.$

GE process as above:

1	0	1	1	interchange rows	Changes the sign of the determinant.
0	1	1	1		
1	1	0	1		
1	1	1	0		

$$\begin{bmatrix} + & - & + & - \\ - & + & - & + \\ + & - & + & - \\ - & + & - & + \end{bmatrix}.$$

Larger determinants are defined similarly.

The above rules hold for determinants of all sizes. Indeed, corresponding results also hold for columns and column operations in all determinants, but we shall not go into the details of these. (Elementary column operations are exactly analogous to elementary row operations.)

It is apparent that evaluation of large determinants will be a lengthy business. The results of this chapter can be used to provide a short cut, however. If we apply the standard GE process, keeping track of all the row operations used, we end up with a triangular matrix, whose determinant is a multiple of the given determinant. Evaluation of determinants of triangular matrices is a simple matter (see Example 7.8), so here is another use for our GE process. Some determinants are evaluated by this procedure in Examples 7.9.

Now recall the Equivalence Theorem from Chapter 6. Four different sets of circumstances led to the GE process applied to a $p \times p$ matrix ending with a matrix with p non-zero rows. The argument above demonstrates that in such a case the $p \times p$ matrix concerned must have a non-zero determinant. The upper triangular matrix resulting from the GE process applied to a $p \times p$ matrix has determinant zero if and only if its last row consists entirely of 0s, in which case it has fewer than p non-zero rows.

Here then is the complete Equivalence Theorem.

Theorem

Let A be a $p \times p$ matrix. The following are equivalent.

(i) A is invertible.

(ii) The rank of A is equal to p.

(iii) The columns of A form a LI list.

(iv) The set of simultaneous equations which can be written $A\boldsymbol{x}=\mathbf{0}$ has no solution other than $\boldsymbol{x}=0$.

(v) (This is the new part.) The determinant of A is non-zero.

We end this chapter with some new notions, which are developed in further study of linear algebra (but not in this book).

In the expansion of a determinant by a row or column, each entry of the chosen row or column is multiplied by a *signed* determinant of the next smaller size. This signed determinant is called the *cofactor* of that particular

$$\begin{array}{rrrrl} 1 & 0 & 1 & 1 & \\ 0 & 1 & 1 & 1 & \\ 0 & 1 & -1 & 0 & (3)-(1) \\ 0 & 1 & 0 & -1 & (4)-(1) \end{array}$$

Determinant unchanged

$$\begin{array}{rrrrl} 1 & 0 & 1 & 1 & \\ 0 & 1 & 1 & 1 & \\ 0 & 0 & -2 & -1 & (3)-(2) \\ 0 & 0 & -1 & -2 & (4)-(2) \end{array}$$

Determinant unchanged.

$$\begin{array}{rrrrl} 1 & 0 & 1 & 1 & \\ 0 & 1 & 1 & 1 & \\ 0 & 0 & 1 & \frac{1}{2} & (3)\div -2 \\ 0 & 0 & -1 & -2 & \end{array}$$

Divides the determinant by -2.

$$\begin{array}{rrrrl} 1 & 0 & 1 & 1 & \\ 0 & 1 & 1 & 1 & \\ 0 & 0 & 1 & \frac{1}{2} & \\ 0 & 0 & 0 & -\frac{3}{2} & (4)+(3) \end{array}$$

Determinant unchanged.

This last matrix has determinant equal to $-\frac{3}{2}$. Hence the original determinant has value $(-\frac{3}{2})\times(-2)$, i.e. 3.

7.10 Find adj A, where

$$A=\begin{bmatrix} 2 & 1 & 1 \\ -1 & 1 & 0 \\ 1 & -1 & 1 \end{bmatrix}.$$

$$A_{11}=\begin{vmatrix} 1 & 0 \\ -1 & 1 \end{vmatrix}=1, \qquad A_{12}=-\begin{vmatrix} -1 & 0 \\ 1 & 1 \end{vmatrix}=1,$$

$$A_{13}=\begin{vmatrix} -1 & 1 \\ 1 & -1 \end{vmatrix}=0, \qquad A_{21}=-\begin{vmatrix} 1 & 1 \\ -1 & 1 \end{vmatrix}=-2,$$

$$A_{22}=\begin{vmatrix} 2 & 1 \\ 1 & 1 \end{vmatrix}=1, \qquad A_{23}=-\begin{vmatrix} 2 & 1 \\ 1 & -1 \end{vmatrix}=3,$$

$$A_{31}=\begin{vmatrix} 1 & 1 \\ 1 & 0 \end{vmatrix}=-1, \qquad A_{32}=-\begin{vmatrix} 2 & 1 \\ -1 & 0 \end{vmatrix}=-1,$$

$$A_{33}=\begin{vmatrix} 2 & 1 \\ -1 & 1 \end{vmatrix}=3.$$

So

$$\text{adj}\,A=\begin{bmatrix} A_{11} & A_{21} & A_{31} \\ A_{12} & A_{22} & A_{32} \\ A_{13} & A_{23} & A_{33} \end{bmatrix}=\begin{bmatrix} 1 & -2 & -1 \\ 1 & 1 & -1 \\ 0 & 3 & 3 \end{bmatrix}.$$

entry. We illustrate this in the 3×3 case, but the following ideas apply to determinants of any size. In the determinant

$$\begin{vmatrix} a_1 & a_2 & a_3 \\ b_1 & b_2 & b_3 \\ c_1 & c_2 & c_3 \end{vmatrix},$$

the cofactor of a_1 is

$$\begin{vmatrix} b_2 & b_3 \\ c_2 & c_3 \end{vmatrix},$$

the cofactor of a_2 is

$$-\begin{vmatrix} b_1 & b_3 \\ c_1 & c_3 \end{vmatrix},$$

and so on. If we change notation to a double suffix and let

$$A=\begin{bmatrix} a_{11} & a_{12} & a_{13} \\ a_{21} & a_{22} & a_{23} \\ a_{31} & a_{32} & a_{33} \end{bmatrix},$$

then there is a convenient notation for cofactors. The cofactor of a_{ij} is denoted by A_{ij}. For example:

$$A_{11}=\begin{vmatrix} a_{22} & a_{23} \\ a_{32} & a_{33} \end{vmatrix} \quad \text{and} \quad A_{12}=-\begin{vmatrix} a_{21} & a_{23} \\ a_{31} & a_{33} \end{vmatrix}.$$

In this notation, then, we can write

$$\det A=a_{11}A_{11}+a_{12}A_{12}+a_{13}A_{13},$$

which is the expansion by the first row, and similarly for expansions by the other rows and the columns. Note that the negative signs are incorporated in the cofactors, so the determinant is represented in this way as a *sum* of terms.

The *adjoint* matrix of A (written adj A) is defined as follows. The (i,j)-entry of adj A is A_{ji}. Note the order of the suffixes. To obtain adj A from A, replace each entry a_{ij} of A by its own cofactor A_{ij}, and then transpose the resulting matrix. This yields adj A. See Example 7.10. This process is impossibly long in practice, even for matrices as small as 3×3, so the significance of the adjoint matrix is mainly theoretical. We can write down one (perhaps surprising) result, however.

Theorem

If A is an invertible matrix, then

$$A^{-1}=\frac{1}{\det A}\,\text{adj}\,A.$$

7.11 Show that if A is an invertible 3×3 matrix then

$$A^{-1}=1/(\det A)\,\text{adj}\,A.$$

Let $A=[a_{ij}]_{3\times 3}$. The adjoint of A is the *transposed* matrix of cofactors, so the (k,j)-entry in adj A is A_{jk}. Hence the (i,j)-entry in the product $A(\text{adj}\,A)$ is

$$\sum_{k=1}^{3} a_{ik}A_{jk},\quad \text{i.e } a_{i1}A_{j1}+a_{i2}A_{j2}+a_{i3}A_{j2}\quad (*).$$

Now if $j=i$ then this is equal to

$$a_{i1}A_{i1}+a_{i2}A_{i2}+a_{i3}A_{i3},$$

which is the value of det A (expanded by the ith row). So every entry of $A(\text{adj}\,A)$ on the main diagonal is det A. Moreover, if $j\neq i$, then the (i,j)-entry in $A(\text{adj}\,A)$ is zero. This is because the expression (*) then is in effect the expansion of a determinant in which two rows are identical. For example, if $i=2$ and $j=1$:

$$a_{21}A_{11}+a_{22}A_{12}+a_{23}A_{13}=\begin{vmatrix} a_{21} & a_{22} & a_{23}\\ a_{21} & a_{22} & a_{23}\\ a_{31} & a_{32} & a_{33}\end{vmatrix}=0.$$

Hence all entries in $A(\text{adj}\,A)$ which are off the main diagonal are zero. So

$$A(\text{adj}\,A)=\begin{bmatrix} \det A & 0 & 0\\ 0 & \det A & 0\\ 0 & 0 & \det A\end{bmatrix}=(\det A)I.$$

It follows that

$$A(1/(\det A)\,\text{adj}\,A)=I.$$

(The supposition that A is invertible ensures that det $A\neq 0$.)

Consequently (see Example 5.10), we have the required result:

$$A^{-1}=1/(\det A)\,\text{adj}\,A.$$

The above argument can be extended to deal with the case of a $p\times p$ matrix, for any value of p.

Proof: See Example 7.11. See also Example 5.8.

Finally, an important property of determinants. We omit the proof, which is not easy. The interested reader may find a proof in the book by Kolman. (A list of books for further reading is given on page 146.)

Theorem

If A and B are $p\times p$ matrices, then

$$\det(AB)=(\det A)(\det B).$$

Summary

Definitions are given of 2×2 and 3×3 determinants, and methods are described for evaluating such determinants. It is shown how larger determinants can be defined and evaluated. The effects on determinants of elementary row operations are shown. The application of the GE process to evaluating determinants is demonstrated, and it is used to show that a square matrix is invertible if and only if it has a non-zero determinant. Cofactors and the adjoint matrix are defined. The theorem on the determinant of a product of matrices is stated.

Exercises

1. Evaluate the following determinants.

$$\begin{vmatrix}2 & 1\\ 3 & -2\end{vmatrix},\quad \begin{vmatrix}-1 & -2\\ -3 & -4\end{vmatrix},\quad \begin{vmatrix}3 & -5\\ 0 & 1\end{vmatrix},\quad \begin{vmatrix}4 & -2\\ 0 & 0\end{vmatrix},$$

$$\begin{vmatrix}4 & 2\\ 6 & -4\end{vmatrix},\quad \begin{vmatrix}-2 & -1\\ -3 & 2\end{vmatrix},\quad \begin{vmatrix}3 & -2\\ 2 & 1\end{vmatrix},\quad \begin{vmatrix}0 & 1\\ 3 & -5\end{vmatrix}.$$

2. Evaluate the following determinants (by any of the procedures described in the text for 3×3 determinants).

$$\begin{vmatrix}0 & 1 & 1\\ 1 & 0 & 1\\ 1 & 1 & 0\end{vmatrix},\quad \begin{vmatrix}3 & 1 & 3\\ -2 & -1 & 0\\ 1 & 1 & 1\end{vmatrix},\quad \begin{vmatrix}0 & 0 & 1\\ -2 & 1 & 2\\ 1 & 4 & -6\end{vmatrix},$$

$$\begin{vmatrix}1 & 2 & 3\\ 4 & 5 & 6\\ 7 & 8 & 9\end{vmatrix},\quad \begin{vmatrix}4 & 5 & 1\\ 1 & 1 & -1\\ 3 & 2 & 2\end{vmatrix},\quad \begin{vmatrix}1 & 1 & -1\\ 4 & 5 & 1\\ 3 & 2 & 2\end{vmatrix},$$

$$\begin{vmatrix}2 & 4 & -6\\ -1 & -2 & 3\\ 1 & 1 & 4\end{vmatrix},\quad \begin{vmatrix}3 & 0 & 3\\ -1 & 1 & 1\\ -2 & 0 & -2\end{vmatrix},\quad \begin{vmatrix}2 & -2 & 3\\ 1 & 0 & 4\\ 0 & 1 & 1\end{vmatrix}.$$

3. Evaluate the following 4×4 determinants.

$$\begin{vmatrix} 2 & 0 & 1 & -1 \\ -1 & 1 & 0 & 3 \\ 0 & 2 & 1 & 1 \\ 3 & 3 & 1 & -1 \end{vmatrix}, \quad \begin{vmatrix} 0 & 3 & 1 & -2 \\ 2 & 2 & -2 & 1 \\ 1 & 0 & 1 & 0 \\ 0 & 2 & -3 & 3 \end{vmatrix}.$$

4. Let A be a 3×3 skew-symmetric matrix. Prove that $\det A = 0$. Is this true for all skew-symmetric matrices?

5. Using the fact that, for any square matrices A and B of the same size, $\det(AB) = (\det A)(\det B)$, show that if either A or B is singular (or both are) then AB is singular. Show also that if A is invertible then $\det(A^{-1}) = 1/(\det A)$.

6. Let A be a 3×3 matrix, and let k be any real number. Show that $\det(kA) = k^3(\det A)$.

7. Let A be a square matrix such that $A^3 = 0$. Show that $\det A = 0$, so that A is singular. Extend this to show that every square matrix A which satisfies $A^n = 0$, for some n, is singular.

8. Evaluate $\det A$ and $\operatorname{adj} A$, where

$$A = \begin{bmatrix} 0 & 1 & 1 \\ 1 & 0 & 1 \\ 1 & 1 & 0 \end{bmatrix}.$$

Check your answers by evaluating the product $A(\operatorname{adj} A)$.

Examples

8.1 By finding the ranks of appropriate matrices, decide whether the following set of equations has any solutions.

$$x_1+3x_2+2x_3=\ \ 3$$
$$-x_1+\ x_2+2x_3=-3$$
$$2x_1+4x_2-2x_3=\ 10.$$

$$A=\begin{bmatrix}1&3&2\\-1&1&2\\2&4&-2\end{bmatrix},\quad [A\,\vdots\,\boldsymbol{h}]=\begin{bmatrix}1&3&2&3\\-1&1&2&-3\\2&4&-2&10\end{bmatrix}.$$

The GE process yields (respectively)

$$\begin{bmatrix}1&3&2\\0&1&1\\0&0&1\end{bmatrix}\qquad\begin{bmatrix}1&3&2&3\\0&1&1&0\\0&0&1&-1\end{bmatrix}$$

The rank of A is 3, the rank of $[A\,\vdots\,\boldsymbol{h}]$ is 3, so there do exist solutions.

8.2 Examples of ranks of augmented matrices.

(i) $$[A\,\vdots\,\boldsymbol{h}]=\begin{bmatrix}0&1&2&-3\\1&-1&0&-2\\3&-2&2&1\end{bmatrix}.$$

The GE process leads to

$$\begin{bmatrix}1&-1&0&-2\\0&1&2&-3\\0&0&0&1\end{bmatrix},$$

so the rank of A is 2 and the rank of $[A\,\vdots\,\boldsymbol{h}]$ is 3.

(ii) (Arising from a set of four equations in three unknowns.)

$$[A\,\vdots\,\boldsymbol{h}]=\begin{bmatrix}1&-1&1&4\\0&2&2&6\\2&3&-1&8\\-1&2&0&-1\end{bmatrix}.$$

The GE process leads to

$$\begin{bmatrix}1&-1&1&4\\0&1&1&3\\0&0&1&\frac{15}{8}\\0&0&0&0\end{bmatrix},$$

so the rank of A is 3 and the rank of $[A\,\vdots\,\boldsymbol{h}]$ is 3.

(iii) (Arising from a set of four equations in three unknowns.)

$$[A\,\vdots\,\boldsymbol{h}]=\begin{bmatrix}1&1&1&1\\3&4&-1&-2\\-1&0&2&1\\0&2&1&0\end{bmatrix}.$$

The GE process leads to

$$\begin{bmatrix}1&1&1&1\\0&1&-4&-5\\0&0&1&1\\0&0&0&1\end{bmatrix}.$$

8

Solutions to simultaneous equations 2

We have developed ideas since Chapter 2 which are applicable to solving simultaneous linear equations, so let us reconsider our methods in the light of these ideas. Recall that a set of p equations in q unknowns can be written in the form $A\boldsymbol{x}=\boldsymbol{h}$, where A is the $p\times q$ matrix of coefficients, $\boldsymbol{x}$ is the q-vector of unknowns and $\boldsymbol{h}$ is the p-vector of the right-hand sides of the equations. Recall also that there can be three possible situations: no solution, a unique solution or infinitely many solutions.

Example 8.1 illustrates the criterion for deciding whether there are any solutions. Let $[A\,|\,\boldsymbol{h}]$ denote the *augmented matrix* obtained by adding $\boldsymbol{h}$ as an extra column to A ($[A\,|\,\boldsymbol{h}]$ is the matrix on which we carry out the GE process). As we saw in Chapter 2, the equations are *in*consistent if and only if the last non-zero row (after the GE process) consists entirely of 0s except for the last entry. Consider what this means with regard to the ranks of the matrices A and $[A\,|\,\boldsymbol{h}]$. The GE process applied to $[A\,|\,\boldsymbol{h}]$ is identical to the GE process applied to A, as far as the first q columns are concerned. In the above situation, then, the rank of A is less than the rank of $[A\,|\,\boldsymbol{h}]$, since the last non-zero row after the GE process on $[A\,|\,\boldsymbol{h}]$ corresponds to a row of 0s in the matrix obtained from A by the GE process. We therefore have:

Rule

The equation $A\boldsymbol{x}=\boldsymbol{h}$ has a solution if and only if the rank of $[A\,|\,\boldsymbol{h}]$ is the same as the rank of A.

Examples 8.2 provide illustration of the different cases which arise.

Notice the special case of *homogeneous* simultaneous equations, that is, the case when $\boldsymbol{h}=\boldsymbol{0}$. As we observed before, such a set of equations *must* be consistent, because a solution is obtained by taking every unknown to be zero. A moment's thought should convince the reader that here the rank of $[A\,|\,\boldsymbol{h}]$ is bound to be the same as the rank of A.

Here the rank of A is 3 and the rank of $[A \mid \boldsymbol{h}]$ is 4, so the set of equations would have been inconsistent.

8.3 Illustrations of the equation $A\boldsymbol{x}=\boldsymbol{h}$, with A singular.

(i) $$[A \mid \boldsymbol{h}]=\begin{bmatrix} 1 & 0 & 3 & 5 \\ -2 & 5 & -1 & 0 \\ -1 & 4 & 1 & 4 \end{bmatrix}.$$

The GE process applied to A:

$$\begin{bmatrix} 1 & 0 & 3 \\ -2 & 5 & -1 \\ -1 & 4 & 1 \end{bmatrix} \rightarrow \begin{bmatrix} 1 & 0 & 3 \\ 0 & 1 & 1 \\ 0 & 0 & 0 \end{bmatrix} \quad (A \text{ is singular}).$$

The GE process applied to $[A \mid \boldsymbol{h}]$:

$$\begin{bmatrix} 1 & 0 & 3 & 5 \\ -2 & 5 & -1 & 0 \\ -1 & 4 & 1 & 4 \end{bmatrix} \rightarrow \begin{bmatrix} 1 & 0 & 3 & 5 \\ 0 & 1 & 1 & 2 \\ 0 & 0 & 0 & 1 \end{bmatrix}.$$

In this case there would be no solutions, since the ranks of A and $[A \mid \boldsymbol{h}]$ are unequal.

(ii) $$[A \mid \boldsymbol{h}]=\begin{bmatrix} 1 & -1 & 3 & -4 \\ 2 & 3 & 1 & 7 \\ 4 & 3 & 5 & 5 \end{bmatrix}.$$

The GE process applied to A:

$$\begin{bmatrix} 1 & -1 & 3 \\ 2 & 3 & 1 \\ 4 & 3 & 5 \end{bmatrix} \rightarrow \begin{bmatrix} 1 & -1 & 3 \\ 0 & 1 & -1 \\ 0 & 0 & 0 \end{bmatrix} \quad (A \text{ is singular}).$$

The GE process applied to $[A \mid \boldsymbol{h}]$:

$$\begin{bmatrix} 1 & -1 & 3 & -4 \\ 2 & 3 & 1 & 7 \\ 4 & 3 & 5 & 5 \end{bmatrix} \rightarrow \begin{bmatrix} 1 & -1 & 3 & -4 \\ 0 & 1 & -1 & 3 \\ 0 & 0 & 0 & 0 \end{bmatrix}.$$

In this case there would be infinitely many solutions. The ranks of A and of $[A \mid \boldsymbol{h}]$ are the same, but less than 3.

8.4 Solution involving two parameters.

$$[A \mid \boldsymbol{h}]=\begin{bmatrix} 1 & -1 & 3 & -2 \\ 2 & -2 & 6 & -4 \\ -1 & 1 & 3 & 2 \end{bmatrix}.$$

The GE process applied to $[A \mid \boldsymbol{h}]$ leads to

$$\begin{bmatrix} 1 & -1 & 3 & -2 \\ 0 & 0 & 0 & 0 \\ 0 & 0 & 0 & 0 \end{bmatrix}.$$

To solve the matrix equation $\begin{bmatrix} x_1 \\ x_2 \\ x_3 \end{bmatrix} = \begin{bmatrix} -2 \\ -4 \\ 2 \end{bmatrix}$ we have, in effect, only the single equation

$$x_1-x_2+3x_3=-2.$$

Introduce parameters $x_2=t$ and $x_3=u$, and substitute to obtain $x_1=t-3u-2$ (the method of Chapter 2).

Next, given a set of equations which are known to have a solution, what criterion determines whether there is a unique solution or infinitely many? Part of the answer is very easy to see.

Rule

If A is an invertible matrix then the equation $A\boldsymbol{x}=\boldsymbol{h}$ has a unique solution. (The solution is $\boldsymbol{x}=A^{-1}\boldsymbol{h}$.)

The other part of the answer is the converse of this, namely:

Rule

If A is a singular matrix then the equation $A\boldsymbol{x}=\boldsymbol{h}$ (if it has solutions at all) has infinitely many solutions.

Proof: See Examples 8.3 for illustration. We must consider the GE process and the process for inverting a matrix. If A is singular, then the GE process applied to A yields a matrix whose last row consists entirely of 0s. The GE process applied to $[A \mid \boldsymbol{h}]$ may have last row all 0s or may have last row all 0s except for the last entry. In the latter case there are no solutions, and in the former case we have to look at the last non-zero row in order to decide whether there are no solutions, or infinitely many solutions. See the rule given in Chapter 2.

Example 8.4 shows a particularly trivial way in which there can be infinitely many solutions. In that case there are two parameters in the solution. Example 8.5 (in which the matrix is 4×4) shows that two parameters can arise in the solution of non-trivial cases also. Can you suggest a 4×4 set of equations in which the set of solutions has three parameters? All these are examples of a general rule.

Rule

If A is a $p\times p$ matrix whose rank is r, and $\boldsymbol{h}$ is a p-vector, then the equation $A\boldsymbol{x}=\boldsymbol{h}$ has solutions provided that the rank of $[A \mid \boldsymbol{h}]$ is also equal to r, and in that case the number of parameters needed to specify the solutions is $p-r$. (This covers the case when $r=p$, A is invertible and there is a unique solution which requires no parameters.)

A proof of this rule is beyond the scope of this book, but the reader should be able to see intuitively why it happens by visualising the possible outcomes of the GE process.

Rule

If A is a $p\times q$ matrix with $p\geqslant q$, and $\boldsymbol{h}$ is a p-vector, then the equation $A\boldsymbol{x}=\boldsymbol{h}$ has a unique solution if and only if the rank of A and the rank of $[A \mid \boldsymbol{h}]$ are both equal to q, the number of columns of A.

8.5 Solution involving two parameters.

$$[A \mid h] = \begin{bmatrix} 1 & 2 & 1 & 0 & 1 \\ 1 & -1 & -2 & 3 & -2 \\ -2 & 1 & 3 & -5 & 3 \\ 0 & 2 & 2 & -2 & 2 \end{bmatrix}.$$

The GE process applied to $[A \mid h]$ leads to

$$\begin{bmatrix} 1 & 2 & 1 & 0 & 1 \\ 0 & 1 & 1 & -1 & 1 \\ 0 & 0 & 0 & 0 & 0 \\ 0 & 0 & 0 & 0 & 0 \end{bmatrix}.$$

Solving equations in this case would entail solving *two* equations in four unknowns:

$$x_1 + 2x_2 + x_3 \qquad = 1$$

$$x_2 + x_3 - x_4 = 1.$$

Introduce parameters $x_3 = t$, and $x_4 = u$, and substitute to obtain $x_2 = 1 - t - u$ and $x_1 = -1 + t - 2u$.

8.6 Uniqueness of solutions when A is not square. Listed below are four possible results of applying the GE process (to the augmented matrix) where A is a 4×3 matrix and h is a 4-vector.

(i) $$\begin{bmatrix} 1 & -1 & 2 & 1 \\ 0 & 1 & 3 & -3 \\ 0 & 0 & 1 & -1 \\ 0 & 0 & 0 & 0 \end{bmatrix}.$$

Here the ranks of A and of $[A \mid h]$ are both 3, and there is a unique solution.

(ii) $$\begin{bmatrix} 1 & 0 & 1 & -3 \\ 0 & 1 & 1 & 0 \\ 0 & 0 & 1 & 2 \\ 0 & 0 & 0 & 1 \end{bmatrix}.$$

Here the ranks of A and of $[A \mid h]$ are different, so there are no solutions.

(iii) $$\begin{bmatrix} 1 & 1 & -2 & 2 \\ 0 & 1 & 1 & 3 \\ 0 & 0 & 0 & 1 \\ 0 & 0 & 0 & 0 \end{bmatrix}.$$

Here the ranks of A and of $[A \mid h]$ are different, so there are no solutions.

(iv) $$\begin{bmatrix} 1 & 0 & 3 & -2 \\ 0 & 0 & 1 & 2 \\ 0 & 0 & 0 & 0 \\ 0 & 0 & 0 & 0 \end{bmatrix}.$$

Here the ranks of A and of $[A \mid h]$ are both 2, so there are infinitely many solutions.

To see this, consider what must be the result of the GE process if there is to be a unique solution. The first q rows of the matrix must have 1s in the (1, 1)-, (2, 2)-, ..., (q, q)-places and 0s below this diagonal, and all subsequent rows must consist of 0s only. This is just the situation when both A and $[A \mid \boldsymbol{h}]$ have rank equal to q. Example 8.6 illustrates this.

Notice that if A is a $p \times q$ matrix with $p<q$, then the equation $A\boldsymbol{x}=\boldsymbol{h}$ *cannot* have a unique solution.

Summary

Rules are given and discussed regarding the solution of equations of the form $A\boldsymbol{x}=\boldsymbol{h}$. These involve the rank of A and the rank of the augmented matrix $[A \mid \boldsymbol{h}]$, and whether (in the case where A is a square matrix) A is invertible or singular.

Exercises

By considering the ranks of the matrix of coefficients and the augmented matrix, decide in each case below whether the given set of equations is consistent or not and, if it is, whether there is a unique solution.

(i) $$\begin{aligned} 2x - y &= 1 \\ x + 3y &= 11. \end{aligned}$$

(ii) $$\begin{aligned} 3x - 6y &= 5 \\ x - 2y &= 1. \end{aligned}$$

(iii) $$\begin{aligned} -4x + 3y &= 0 \\ 12x - 9y &= 0. \end{aligned}$$

(iv) $$\begin{aligned} x + 2y &= 0 \\ 3x - 4y &= 0. \end{aligned}$$

(v) $$\begin{aligned} x_1 - x_2 + 2x_3 &= 3 \\ 2x_1 - 3x_2 - x_3 &= -8 \\ 2x_1 + x_2 + x_3 &= 3. \end{aligned}$$

(vi) $$\begin{aligned} 2x_2 + x_3 &= 0 \\ x_1 - 3x_2 + 2x_3 &= 0 \\ 2x_1 + x_2 - x_3 &= 0. \end{aligned}$$

(vii) $$\begin{aligned} -x_1 + 2x_2 - 4x_3 &= 1 \\ 2x_1 + 3x_2 + x_3 &= -2 \\ x_1 - x_2 + 3x_3 &= 2. \end{aligned}$$

(viii) $$\begin{aligned} x_1 + 2x_2 + 3x_3 &= 0 \\ x_1 - x_2 - 3x_3 &= 0 \\ -3x_1 + x_2 - 5x_3 &= 0. \end{aligned}$$

(ix) $$\begin{aligned} 2x_1 + x_2 + 5x_3 &= 3 \\ -x_1 + 2x_2 &= 1 \\ x_1 + 2x_2 + 4x_3 &= 3. \end{aligned}$$

(x) $$\begin{aligned} x_1 - 2x_2 - x_3 &= 2 \\ x_2 + x_3 &= 0 \\ x_1 + x_3 &= 2 \\ x_1 + 3x_2 + 4x_3 &= 3. \end{aligned}$$

(xi) $$\begin{aligned} x_1 - 2x_2 - x_3 &= 0 \\ x_2 + x_2 &= 0 \\ x_1 + x_3 &= 0 \\ x_1 + 3x_2 + 4x_3 &= 0. \end{aligned}$$

(xii) $$\begin{aligned} x_1 + x_2 - x_3 &= 8 \\ -x_1 + x_2 + x_3 &= 2 \\ x_1 - x_2 + x_3 &= 0 \\ x_1 + x_2 + x_3 &= 10. \end{aligned}$$

Examples

9.1

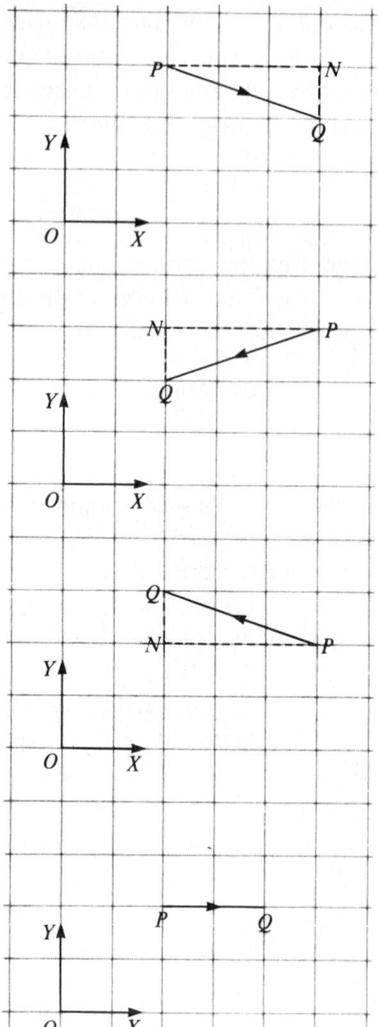

(i) $\overrightarrow{PQ}=\begin{bmatrix} 3 \\ -1 \end{bmatrix}$

(ii) $\overrightarrow{PQ}=\begin{bmatrix} -3 \\ -1 \end{bmatrix}$

(iii) $\overrightarrow{PQ}=\begin{bmatrix} -3 \\ 1 \end{bmatrix}$

(iv) $\overrightarrow{PQ}=\begin{bmatrix} 2 \\ 0 \end{bmatrix}$

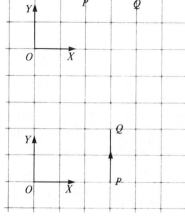

(v) $\overrightarrow{PQ}=\begin{bmatrix} 0 \\ 2 \end{bmatrix}$

9

Vectors in geometry

Linear algebra and geometry are fundamentally related in a way which can be useful in the study of either topic. Ideas from each can provide helpful insights in the other.

The basic idea is that a column vector may be used to represent the position of a point in relation to another point, when coordinate axes are given. This applies in both two-dimensional geometry and three-dimensional geometry, but to start with it will be easier to think of the two-dimensional case. Let P and Q be two (distinct) points, and let OX and OY be given coordinate axes. Draw through P a straight line parallel to OX and through Q a straight line parallel to OY, and let these lines meet at N, as shown.

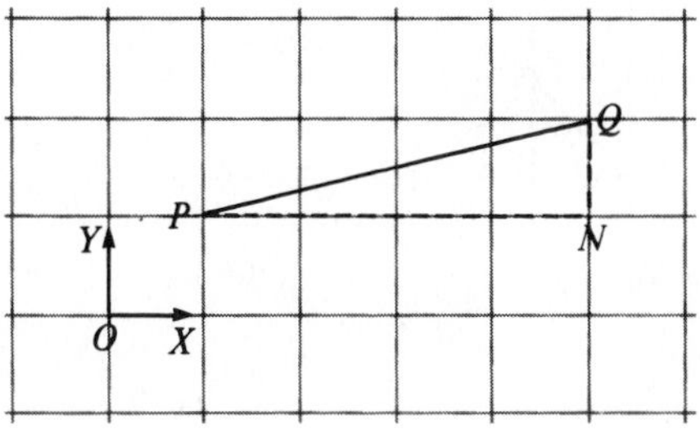

The position of Q relative to P can now be specified by a pair of numbers determined by the lengths and directions of the lines PN and NQ. The sizes of the numbers are the lengths of the lines. The signs of the numbers depend on whether the directions (P to N and N to Q) are the same as or opposite to the directions of the coordinate axes OX and OY respectively.

We can thus associate with the (ordered) pair of points P, Q a column vector $\begin{bmatrix} a \\ b \end{bmatrix}$, where a and b are the two numbers determined by the above

9.2

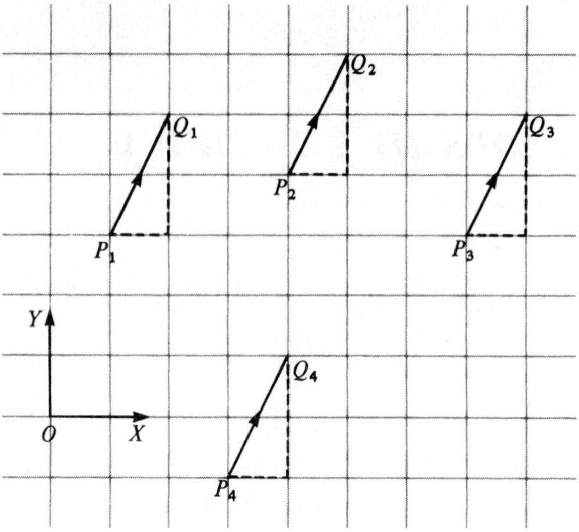

$$\overrightarrow{P_1Q_1}=\begin{bmatrix}2\\1\end{bmatrix},\quad \overrightarrow{P_2Q_2}=\begin{bmatrix}2\\1\end{bmatrix},\quad \overrightarrow{P_3Q_3}=\begin{bmatrix}2\\1\end{bmatrix},\quad \overrightarrow{P_4Q_4}=\begin{bmatrix}2\\1\end{bmatrix}.$$

9.3

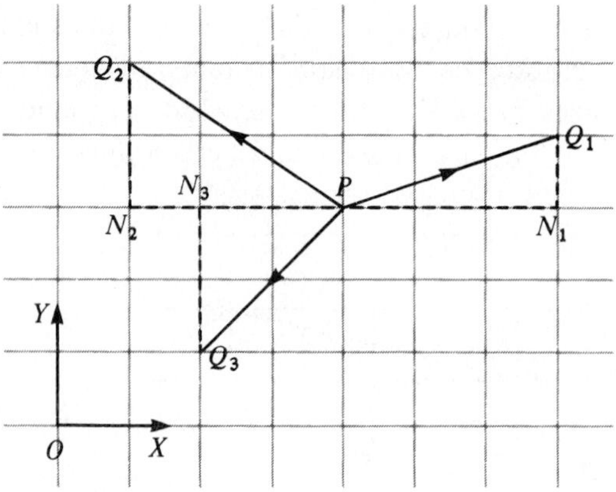

Let $v_1=\begin{bmatrix}3\\1\end{bmatrix}$.

Then Q_1 is the only point such that $\overrightarrow{PQ}_1=v_1$.

Let $v_2=\begin{bmatrix}-3\\2\end{bmatrix}$.

Then Q_2 is the only point such that $\overrightarrow{PQ}_2=v_2$.

Let $v_3=\begin{bmatrix}-2\\-2\end{bmatrix}$.

Then Q_3 is the only point such that $\overrightarrow{PQ}_3=v_3$.

process. The notation $\overrightarrow{PQ}$ is used for this. Examples 9.1 give several pairs of points and their associated column vectors, illustrating the way in which negative numbers can arise. Note the special cases given in Examples 9.1(iv) and (v).

Example 9.2 shows clearly that for any given column vector, many different pairs of points will be associated with it in this way. The diagram shows the properties that the lines P_1Q_1, P_2Q_2, P_3Q_3, and P_4Q_4 have which cause this. They are parallel, they have the same (not opposite) directions, and they have equal lengths.

A column vector is associated with a direction and a length, as we have just seen. Thus, given a (non-zero) column vector $\mathbf{v}$ and a point P, there will always be one and only one point Q such that $\overrightarrow{PQ}=\mathbf{v}$. Example 9.3 illustrates this.

To summarise:

1. Given any points P and Q, there is a unique column vector $\overrightarrow{PQ}$ which represents the position of Q relative to P.
2. Given any non-zero column vector $\mathbf{v}$, there are infinitely many pairs of points P, Q such that $\overrightarrow{PQ}=\mathbf{v}$.
3. Given any point P and any non-zero column vector $\mathbf{v}$, there is a unique point Q such that $\overrightarrow{PQ}=\mathbf{v}$.

9.4

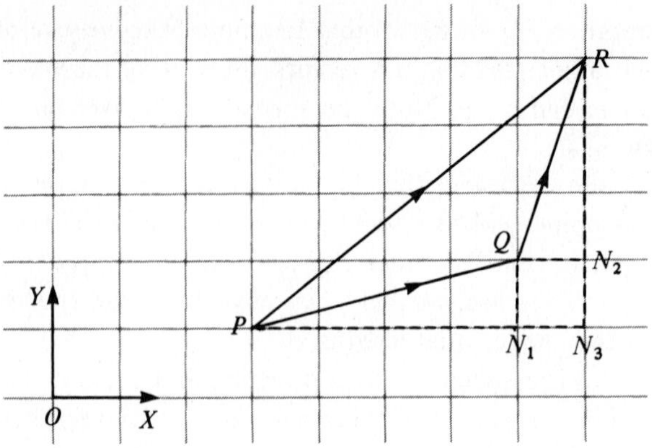

Here we treat only the case when the components of the vectors $\overrightarrow{PQ}$ and $\overrightarrow{QR}$ are all positive. If any are negative, the diagrams will be different and the argument will have to be modified slightly. Let

$$\overrightarrow{PQ} = \begin{bmatrix} a_1 \\ a_2 \end{bmatrix} \quad \text{and} \quad \overrightarrow{QR} = \begin{bmatrix} b_1 \\ b_2 \end{bmatrix}.$$

Then $|PN_1| = a_1$, $|N_1Q| = a_2$, $|QN_2| = b_1$, $|N_2R| = b_2$. So $|PN_3| = |PN_1| + |N_1N_3| = |PN_1| + |QN_2| = a_1 + b_1$, and $|N_3R| = |N_3N_2| + |N_2R| = |N_1Q| + |N_2R| = a_2 + b_2$.

9.5

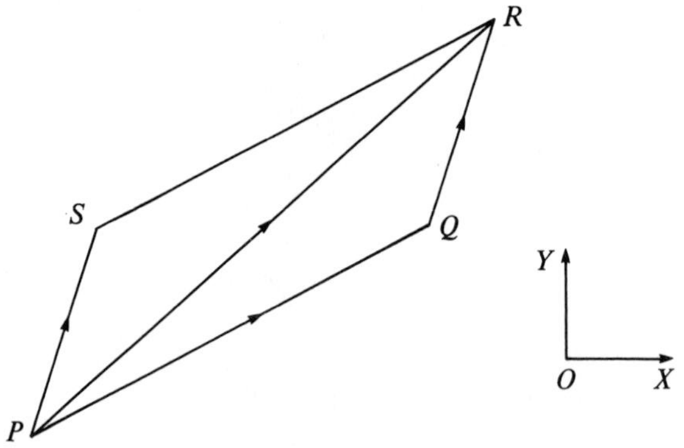

$PQRS$ is a parallelogram. Hence QR and PS have the same direction and the same length. Consequently they are associated with the same column vector, i.e. $\overrightarrow{QR} = \overrightarrow{PS}$. By the Triangle Law, $\overrightarrow{PQ} + \overrightarrow{QR} = \overrightarrow{PR}$, so it follows immediately that $\overrightarrow{PQ} + \overrightarrow{PS} = \overrightarrow{PR}$.

Where is the advantage in this? It is in the way that addition of column vectors corresponds with a geometrical operation. Let P, Q and R be three distinct points, and let

$$\overrightarrow{PQ}=\begin{bmatrix} a_1 \\ a_2 \end{bmatrix} \quad \text{and} \quad \overrightarrow{QR}=\begin{bmatrix} b_1 \\ b_2 \end{bmatrix}.$$

Then

$$\overrightarrow{PR}=\begin{bmatrix} a_1+b_1 \\ a_2+b_2 \end{bmatrix}=\begin{bmatrix} a_1 \\ a_2 \end{bmatrix}+\begin{bmatrix} b_1 \\ b_2 \end{bmatrix}=\overrightarrow{PQ}+\overrightarrow{QR}.$$

This rule is called the Triangle Law, and is of fundamental importance in vector geometry. See Example 9.4 for a partial justification for it.

Rule (The Triangle Law)
If PQR is a triangle then

$$\overrightarrow{PQ}+\overrightarrow{QR}=\overrightarrow{PR}.$$

The Triangle Law is sometimes expressed in a different form:

Rule (The Parallelogram Law)
If $PQRS$ is a parallelogram then

$$\overrightarrow{PQ}+\overrightarrow{PS}=\overrightarrow{PR}.$$

See Example 9.5 for a justification of this, using the Triangle Law.

9.6 Position vectors.

(i)

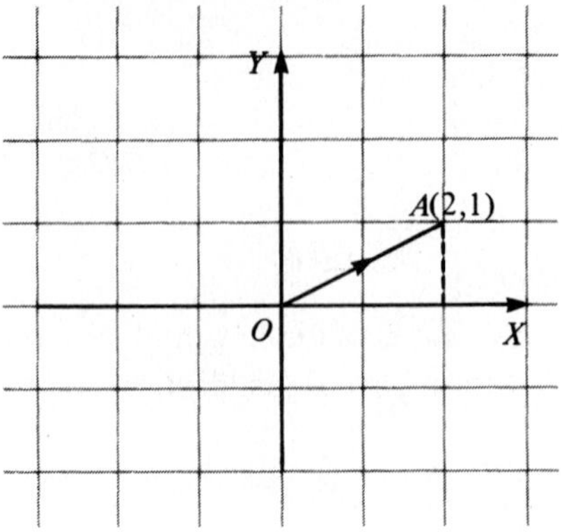

$$\boldsymbol{r}_A = \begin{bmatrix} 2 \\ 1 \end{bmatrix}$$

(ii)

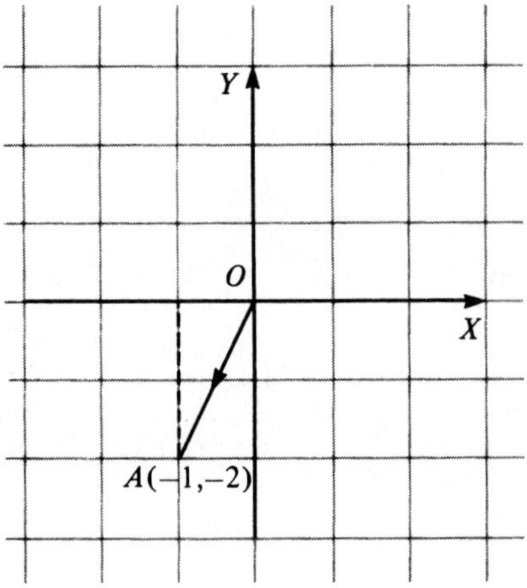

$$\boldsymbol{r}_A = \begin{bmatrix} -1 \\ -2 \end{bmatrix}$$

(iii)

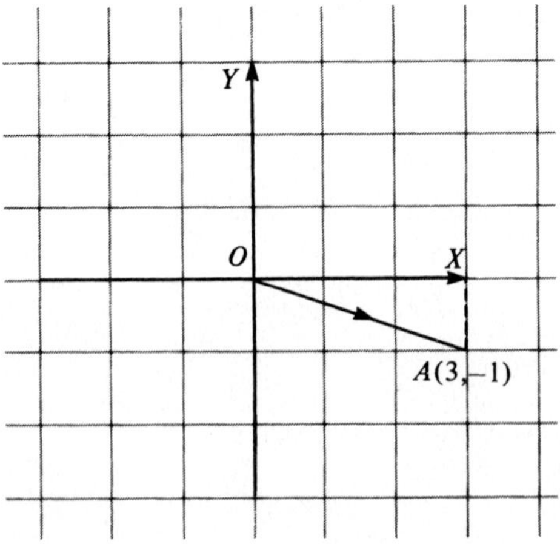

$$\boldsymbol{r}_A = \begin{bmatrix} 3 \\ -1 \end{bmatrix}.$$

This relationship between algebraic vectors and directed lines is dependent on having some origin and coordinate axes, but it is important to realise that these laws which we have found are true irrespective of the choice of coordinate axes. We shall develop ideas and techniques which likewise are geometrical in character but which use the algebra of vectors in a convenient way without the necessity to refer to *particular* coordinate axes.

Before we proceed, we must take note of an important special case of all this, namely when the reference point (the first of the pair) is the origin.

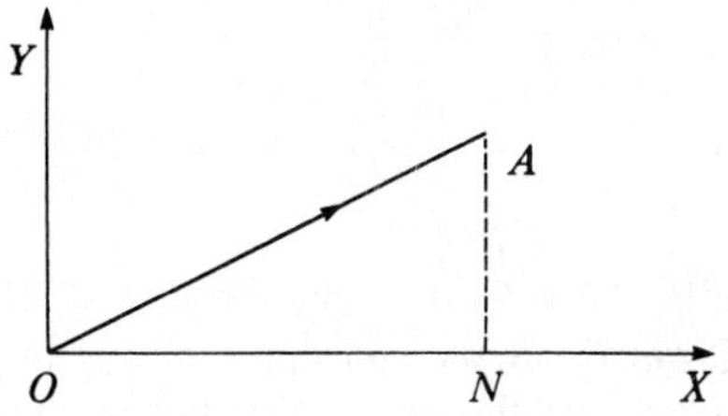

The construction which yields the components of the column vector $\overrightarrow{OA}$ involves the lines ON and NA. But now it is easy to see that the components of $\overrightarrow{OA}$ are just the coordinates of A. See Example 9.6 for particular cases, including cases with negative coordinates. The vector $\overrightarrow{OA}$ is called the *position vector* of A. It is customarily denoted by r_A. Every point has a uniquely determined position vector.

9.7

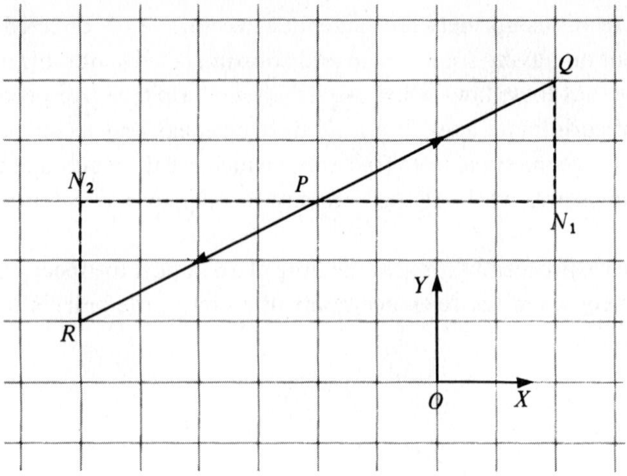

Produce QP back so that $|PR|=|PQ|$. Then the triangles PN_2R and PN_1Q are congruent, so PN_1 and PN_2 have equal lengths and opposite directions. The column vector associated with PR is thus the negative of the column vector associated with PQ. Notice that $\overrightarrow{PQ}+\overrightarrow{PR}=\mathbf{0}$, which can be thought of as an extension of the Parallelogram Law to an extreme case. (The parallelogram is flattened.)

9.8 Multiplication by a scalar.

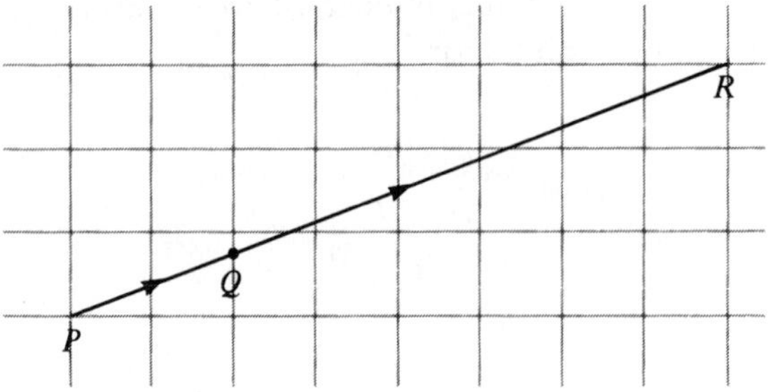

Here $\overrightarrow{PR}=4\,\overrightarrow{PQ}$.

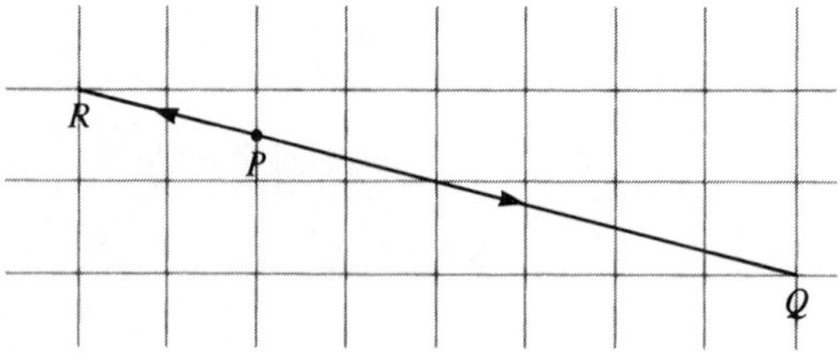

Here $\overrightarrow{PR}=-\frac{1}{3}\,\overrightarrow{PQ}$.

Rules

(i) The zero vector cannot be associated with any pair of distinct points. Nevertheless we can think of **0** as the position vector of the origin itself.

(ii) If $\mathbf{v}$ is a non-zero column vector, and $\overrightarrow{PQ} = \mathbf{v}$, then $-\mathbf{v} = \overrightarrow{PR}$, where R is the point obtained by producing the line QP so that PR and PQ have the same length. See Example 9.7.

(iii) If $\mathbf{v}$ is any non-zero column vector and k is any positive number, and if $\overrightarrow{PQ} = \mathbf{v}$, then $k\mathbf{v} = \overrightarrow{PR}$, where R is the point such that PQ and PR have the same aırection and the length of PR is k times the length of PQ. Multiplying a vector by a negative number reverses the direction (as well as changing the length). See Example 9.8.

9.9 The difference of two vectors. Let PAB be a triangle.

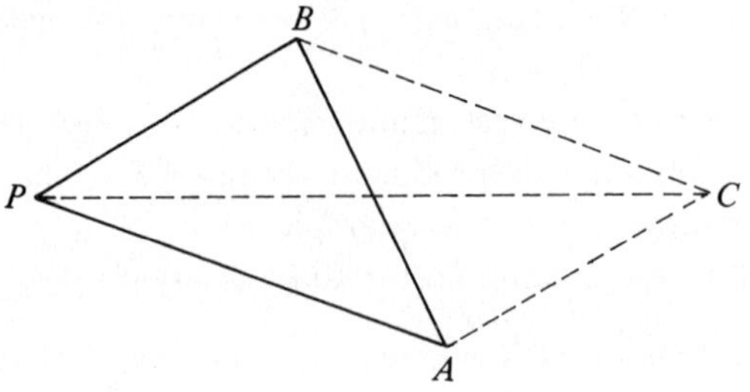

By the Triangle Law,

$$\overrightarrow{PA}+\overrightarrow{AB}=\overrightarrow{PB}.$$

Hence

$$\overrightarrow{AB}=\overrightarrow{PB}-\overrightarrow{PA}.$$

Notice that, in the parallelogram $PACB$, one diagonal (PC) represents the *sum* $\overrightarrow{PA}+\overrightarrow{PB}$, and the other diagonal (AB) represents the *difference* $\overrightarrow{PB}-\overrightarrow{PA}$. Of course the second diagonal may be taken in the opposite direction: $\overrightarrow{BA}=\overrightarrow{PA}-\overrightarrow{PB}$.

9.10 Illustrations of the ratio in which a point divides a line segment. In each case P divides AB in the given ratio.

(i) Ratio 1:1.

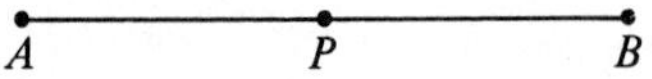

(ii) Ratio 3:1.

(iii) Ratio 2:3.

(iv) Ratio -1:5.

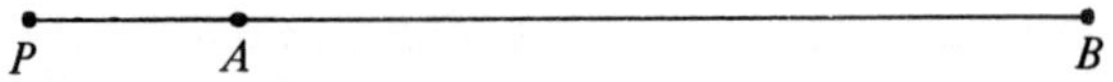

(v) Ratio 4:-1.

Subtraction of vectors also has an important geometrical interpretation. For a diagram, see Example 9.9. In a triangle PAB,

$$\overrightarrow{AB}=\overrightarrow{PB}-\overrightarrow{PA}.$$

This is a consequence of the Triangle Law. In particular, if A and B are any two distinct points then (taking the origin O as the point P) we have

$$\overrightarrow{AB}=\overrightarrow{OB}-\overrightarrow{OA}=\boldsymbol{r}_B-\boldsymbol{r}_A.$$

We shall use this repeatedly.

As mentioned earlier, all of these ideas apply equally well in three dimensions, where points have three coordinates and the vectors associated with directed lines have three components. One of the best features of this subject is the way in which the algebraic properties of 2-vectors and 3-vectors (which are substantially the same) can be used in the substantially different geometrical situations of two and three dimensions.

As an application of algebraic vector methods in geometry we shall derive the *Section Formula*. Consider a line segment AB. A point P on AB (possibly produced) divides AB in the ratio $m:n$ (with m and n both positive) if $|AP|/|PB|=m/n$. Here $|AP|$ and $|PB|$ denote the lengths of the lines AP and PB respectively. Extending this idea, we say that P divides AB in the ratio $-m:n$ (with m and n both positive) if AP and PB have opposite directions and $|AP|/|PB|=m/n$. See Examples 9.10 for illustrations of these.

9.11 Proof of the Section Formula (the case with m and n positive).

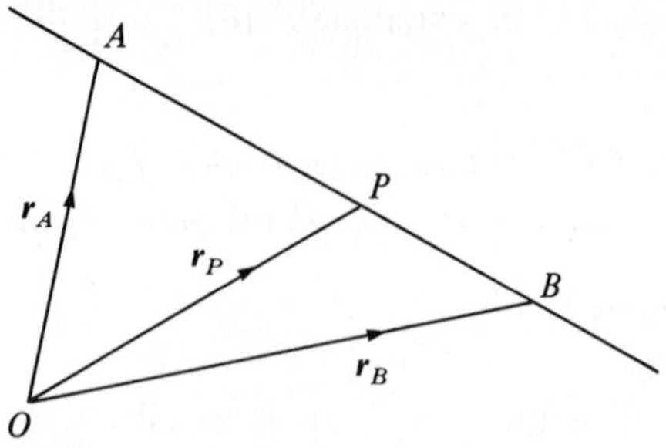

Let P divide AB in the ratio $m{:}n$, with $m>0$ and $n>0$. Then

$$\overrightarrow{AP}=(m/n)\,\overrightarrow{PB}.$$

Now

$$\overrightarrow{AP}=\boldsymbol{r}_P-\boldsymbol{r}_A \quad \text{and} \quad \overrightarrow{PB}=\boldsymbol{r}_B-\boldsymbol{r}_P,$$

so

$$n(\boldsymbol{r}_P-\boldsymbol{r}_A)=m(\boldsymbol{r}_B-\boldsymbol{r}_P).$$

From this we obtain

$$n\boldsymbol{r}_P-n\boldsymbol{r}_A-m\boldsymbol{r}_B+m\boldsymbol{r}_P=\boldsymbol{0},$$

$$(m+n)\boldsymbol{r}_P=n\boldsymbol{r}_A+m\boldsymbol{r}_B,$$

$$\boldsymbol{r}_P=\frac{1}{(m+n)}(n\boldsymbol{r}_A+m\boldsymbol{r}_B).$$

The cases where m and n have opposite signs require separate proofs, but the ideas are the same.

9.12 The medians of a triangle are concurrent at the centroid of the triangle.

Let ABC be a triangle, and let L, M and N be the midpoints of the sides BC, CA and AB, respectively. Let A, B and C have position vectors $\boldsymbol{a}$, $\boldsymbol{b}$ and $\boldsymbol{c}$. Then

L has position vector $\frac{1}{2}(\boldsymbol{b}+\boldsymbol{c})$,

M has position vector $\frac{1}{2}(\boldsymbol{c}+\boldsymbol{a})$, and

N has position vector $\frac{1}{2}(\boldsymbol{a}+\boldsymbol{b})$.

The point which divides AL in the ratio 2:1 has position vector

$$\frac{1}{(2+1)}(1\times\boldsymbol{a}+2\times\tfrac{1}{2}(\boldsymbol{b}+\boldsymbol{c})), \quad \text{i.e. } \tfrac{1}{3}(\boldsymbol{a}+\boldsymbol{b}+\boldsymbol{c}).$$

The point which divides BM in the ratio 2:1 has position vector

$$\frac{1}{(2+1)}(1\times\boldsymbol{b}+2\times\tfrac{1}{2}(\boldsymbol{c}+\boldsymbol{a})), \quad \text{i.e. } \tfrac{1}{3}(\boldsymbol{a}+\boldsymbol{b}+\boldsymbol{c}).$$

The point which divides CN in the ratio 2:1 has position vector

$$\frac{1}{(2+1)}(1\times\boldsymbol{c}+2\times\tfrac{1}{2}(\boldsymbol{a}+\boldsymbol{b})), \quad \text{i.e. } \tfrac{1}{3}(\boldsymbol{a}+\boldsymbol{b}+\boldsymbol{c}).$$

Hence the point with position vector $\frac{1}{3}(\boldsymbol{a}+\boldsymbol{b}+\boldsymbol{c})$ lies on all three medians. This is the centroid of the triangle.

Rule
Let P divide AB in the ratio $m{:}n$ (with $n\neq 0$ and $m+n\neq 0$). Then

$$\boldsymbol{r}_P=\frac{1}{m+n}(n\boldsymbol{r}_A+m\boldsymbol{r}_B).$$

See Example 9.11 for a proof.

An important special case of the Section Formula is when P is the midpoint of AB. In that case

$$\boldsymbol{r}_P=\tfrac{1}{2}(\boldsymbol{r}_A+\boldsymbol{r}_B).$$

The Section Formula may be used to give a convenient proof of a simple geometrical theorem: the medians of a triangle are concurrent at a point which trisects each median. A median of a triangle is a line which joins a vertex of the triangle to the midpoint of the opposite side. Let ABC be any triangle, and let $\boldsymbol{a}$, $\boldsymbol{b}$ and $\boldsymbol{c}$ be position vectors of A, B and C relative to some fixed origin. See Example 9.12. Let L, M and N be the midpoints of BC, CA and AB respectively. Then

$$\overrightarrow{OL}=\tfrac{1}{2}(\boldsymbol{b}+\boldsymbol{c}),\quad \overrightarrow{OM}=\tfrac{1}{2}(\boldsymbol{c}+\boldsymbol{a}),\quad \overrightarrow{ON}=\tfrac{1}{2}(\boldsymbol{a}+\boldsymbol{b}).$$

The Section Formula may now be used to find the position vectors of the points which divide AL, BM and CN respectively each in the ratio 2:1. It turns out to be the same point, which must therefore lie on all three medians. This point is called the *centroid* of the triangle ABC. It has position vector $\frac{1}{3}(\boldsymbol{a}+\boldsymbol{b}+\boldsymbol{c})$.

This result has been demonstrated using vectors in a geometrical style. The link with algebra has not been explicit, except in the algebraic operations on the vectors. Consequently we worked without reference to any particular coordinate system and obtained a purely geometrical result. These ideas can be taken considerably further, but we shall return now to the link with algebra, via coordinates.

9.13 Components of a vector in the directions of the coordinate axes.

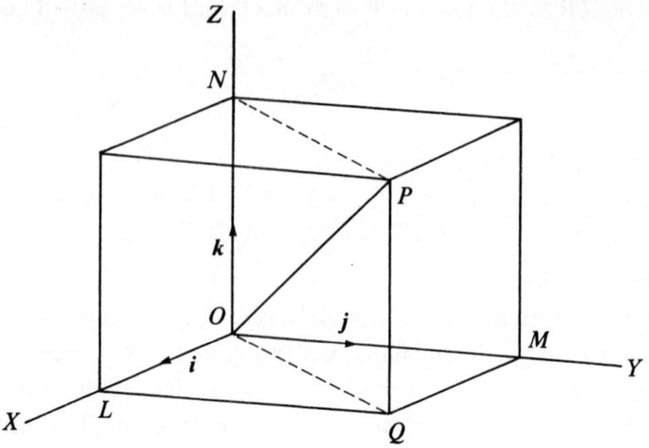

P has coordinates (x, y, z). Construct a rectangular solid figure by drawing perpendiculars from P to the three coordinate planes, and then to the coordinate axes from the feet of these perpendiculars. For example, PQ and then QL and QM, as shown in the diagram. The picture has been simplified by assuming that x, y and z are all positive. Our arguments work also if any or all of them are negative. You should try to visualise the various possible configurations.

$$|OL|=x, \quad |OM|=y, \quad |ON|=z.$$

So

$$\overrightarrow{OL}=x\boldsymbol{i}, \quad \overrightarrow{OM}=y\boldsymbol{j}, \quad \overrightarrow{ON}=z\boldsymbol{k}.$$

Now $\overrightarrow{OL}+\overrightarrow{OM}=\overrightarrow{OQ}$, by the Parallelogram Law. Also $OQPN$ is a parallelogram (it is a rectangle), so $\overrightarrow{OQ}+\overrightarrow{ON}=\overrightarrow{OP}$. Hence

$$\overrightarrow{OP}=\overrightarrow{OL}+\overrightarrow{OM}+\overrightarrow{ON}=x\boldsymbol{i}+y\boldsymbol{j}+z\boldsymbol{k}.$$

9.14 Examples of unit vectors.

(i) Let $\boldsymbol{a}=\begin{bmatrix}1\\2\\1\end{bmatrix}$. Then $\dfrac{1}{|\boldsymbol{a}|}\boldsymbol{a}=\begin{bmatrix}\frac{1}{\sqrt{6}}\\ \frac{2}{\sqrt{6}}\\ \frac{1}{\sqrt{6}}\end{bmatrix}$,

since $|\boldsymbol{a}|=\sqrt{1+4+1}=\sqrt{6}$.

(ii) Let $\boldsymbol{b}=\begin{bmatrix}2\\2\\-1\end{bmatrix}$. Then $\dfrac{1}{|\boldsymbol{b}|}\boldsymbol{b}=\begin{bmatrix}\frac{2}{3}\\ \frac{2}{3}\\ -\frac{1}{3}\end{bmatrix}$,

since $|\boldsymbol{b}|=\sqrt{4+4+1}=3$.

(iii) Let $\boldsymbol{c}=\begin{bmatrix}\frac{3}{5}\\0\\ \frac{4}{5}\end{bmatrix}$. Then $|\boldsymbol{c}|=\sqrt{\frac{9}{25}+0+\frac{16}{25}}=1$.

Thus $\boldsymbol{c}$ is already a unit vector.

Consider three dimensions. A point $P(x, y, z)$ has position vector

$$\boldsymbol{r}_P = \begin{bmatrix} x \\ y \\ z \end{bmatrix}.$$

In particular:

the points $L(1,0,0)$, $M(0,1,0)$ and $N(0,0,1)$ have position vectors

$$\begin{bmatrix} 1 \\ 0 \\ 0 \end{bmatrix}, \quad \begin{bmatrix} 0 \\ 1 \\ 0 \end{bmatrix} \quad \text{and} \quad \begin{bmatrix} 0 \\ 0 \\ 1 \end{bmatrix}$$

respectively.

These three vectors are denoted respectively by $\boldsymbol{i}, \boldsymbol{j}$ and $\boldsymbol{k}$. Note that they are represented by directed lines OL, OM and ON in the directions of the coordinate axes. Notice also that

$$\begin{aligned} \begin{bmatrix} x \\ y \\ z \end{bmatrix} &= \begin{bmatrix} x \\ 0 \\ 0 \end{bmatrix} + \begin{bmatrix} 0 \\ y \\ 0 \end{bmatrix} + \begin{bmatrix} 0 \\ 0 \\ z \end{bmatrix} \\ &= x\begin{bmatrix} 1 \\ 0 \\ 0 \end{bmatrix} + y\begin{bmatrix} 0 \\ 1 \\ 0 \end{bmatrix} + z\begin{bmatrix} 0 \\ 0 \\ 1 \end{bmatrix} \\ &= x\boldsymbol{i} + y\boldsymbol{j} + z\boldsymbol{k}. \end{aligned}$$

The vectors $\boldsymbol{i}, \boldsymbol{j}$ and $\boldsymbol{k}$ are known as *standard basis vectors*. Every vector can be written (uniquely) as a sum of multiples of these, as above. The numbers x, y and z are called the *components* of the vector in the directions of the coordinate axes. See Example 9.13.

It is clear what should be meant by the *length* of a vector: the length of any line segment which represents it. Each of $\boldsymbol{i}, \boldsymbol{j}$ and $\boldsymbol{k}$ above has length equal to 1. A *unit* vector is a vector whose length is equal to 1. The length of an algebraic vector

$$\begin{bmatrix} x \\ y \\ z \end{bmatrix}$$

is (by definition) equal to $\sqrt{x^2+y^2+z^2}$, i.e. the length of the line OP, where P is the point with coordinates (x, y, z). Given any non-zero vector $\boldsymbol{a}$ we can always find a unit vector in the same direction as $\boldsymbol{a}$. If we denote the length of $\boldsymbol{a}$ by $|\boldsymbol{a}|$ then $(1/|\boldsymbol{a}|)\boldsymbol{a}$ is a unit vector in the direction of $\boldsymbol{a}$. See Example 9.14.

Summary

The way in which vectors are represented as directed line segments is explained, and algebraic operations on vectors are interpreted geometrically. The Section Formula is derived and used. The standard basis vectors $\boldsymbol{i}$, $\boldsymbol{j}$ and $\boldsymbol{k}$ are defined. The length of a vector (in three dimensions) is defined, and the notion of a unit vector is introduced.

Exercises

1. Let $ABCDEF$ be a regular hexagon whose centre is at the origin. Let A and B have position vectors $\boldsymbol{a}$ and $\boldsymbol{b}$. Find the position vectors of C, D, E and F in terms of $\boldsymbol{a}$ and $\boldsymbol{b}$.

2. Let $ABCD$ be a parallelogram, and let $\boldsymbol{a}, \boldsymbol{b}, \boldsymbol{c}$ and $\boldsymbol{d}$ be the position vectors of A, B, C and D respectively. Show that $\boldsymbol{a}+\boldsymbol{c}=\boldsymbol{b}+\boldsymbol{d}$.

3. Let L, M and N be the midpoints of BC, CA and AB respectively, and let O be any fixed origin. Show that

   (i) $\overrightarrow{OA}+\overrightarrow{OB}+\overrightarrow{OC}=\overrightarrow{OL}+\overrightarrow{OM}+\overrightarrow{ON}$, and
   (ii) $\overrightarrow{AL}+\overrightarrow{BM}+\overrightarrow{CN}=\boldsymbol{0}$.

4. Let $A_1, A_2, \ldots, A_k$ be any points in three dimensions. Show that

   $\overrightarrow{A_1A_2}+\overrightarrow{A_2A_3}+\cdots+\overrightarrow{A_{k-1}A_k}+\overrightarrow{A_kA_1}=\boldsymbol{0}$.

5. In each case below, write down the 3-vector $\overrightarrow{AB}$, where A and B are points with the given coordinates.

   (i) $A(0,0,0)$, $B(2,-1,3)$.
   (ii) $A(2,-1,3)$, $B(0,0,0)$.
   (iii) $A(3,4,1)$, $B(1,2,-1)$.
   (iv) $A(0,1,-1)$, $B(0,-1,0)$.
   (v) $A(2,2,2)$, $B(3,2,1)$.

6. In each case below, find the position vector of the point which divides the line segment AB in the given ratio.

   (i) $A(1,1,3)$, $B(-1,1,5)$, ratio 1:1.
   (ii) $A(-2,-1,1)$, $B(3,2,2)$, ratio $2:-1$.
   (iii) $A(0,0,0)$, $B(11,11,11)$, ratio 6:5.
   (iv) $A(3,-1,-2)$, $B(10,-8,12)$, ratio $9:-2$.
   (v) $A(2,1,-1)$, $B(-2,-1,1)$, ratio $-2:3$.

   Also in each case draw a rough diagram to indicate the relative positions of the three points.

7. Let $OABC$ be a parallelogram as shown, and let D divide OA in the ratio $m:n$ (where $n\neq 0$ and $m+n\neq 0$). Prove by vector methods that DC and OB intersect at a point P which divides each line in the ratio $m:(m+n)$.

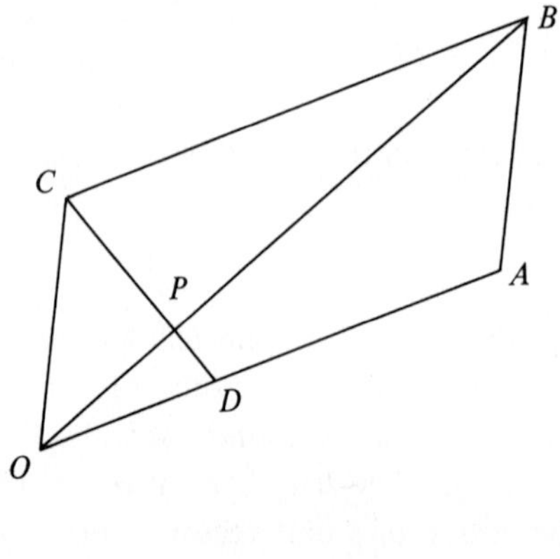

8. Let OAB be a triangle as shown. The midpoint of OA is M, P is the point which divides BM in the ratio 3:2, and S is the point where OP produced meets BA. Prove, using vector methods, that S divides BA in the ratio 3:4.

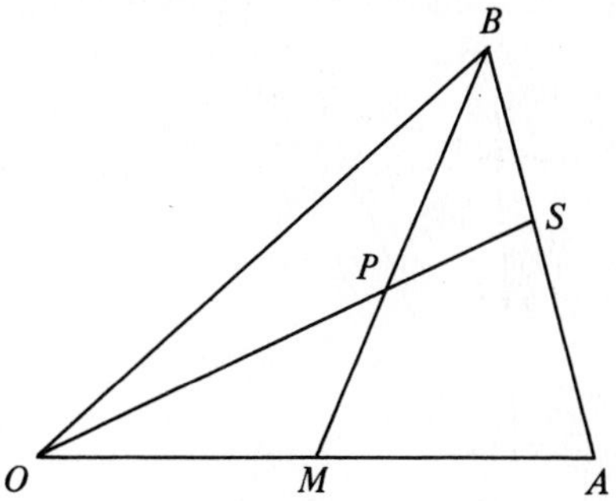

9. In each case below, find a unit vector in the same direction as the given vector.

(i) $\begin{bmatrix} 1 \\ 0 \\ -1 \end{bmatrix}$. (ii) $\begin{bmatrix} 2 \\ 2 \\ 1 \end{bmatrix}$. (iii) $\begin{bmatrix} 1 \\ -2 \\ -1 \end{bmatrix}$. (iv) $\begin{bmatrix} 2 \\ 2 \\ 2 \end{bmatrix}$.

10. Let $\boldsymbol{a}$ and $\boldsymbol{b}$ be non-zero vectors. Prove that $|\boldsymbol{a}+\boldsymbol{b}|=|\boldsymbol{a}|+|\boldsymbol{b}|$ if and only if $\boldsymbol{a}$ and $\boldsymbol{b}$ have the same (or opposite) directions.

11. Let A and B be points with position vectors $\boldsymbol{a}$ and $\boldsymbol{b}$ respectively (with $\boldsymbol{a} \neq \boldsymbol{0}$ and $\boldsymbol{b} \neq \boldsymbol{0}$). Show that $|\boldsymbol{a}|\boldsymbol{b}$ and $|\boldsymbol{b}|\boldsymbol{a}$ are vectors with the same length, and deduce that the direction of the internal bisector of the angle AOB is given by the vector $|\boldsymbol{a}|\boldsymbol{b}+|\boldsymbol{b}|\boldsymbol{a}$.

Examples

10.1 Vector equation of a straight line.

(i) Given two points A and B with position vectors $\boldsymbol{a}$ and $\boldsymbol{b}$ respectively,

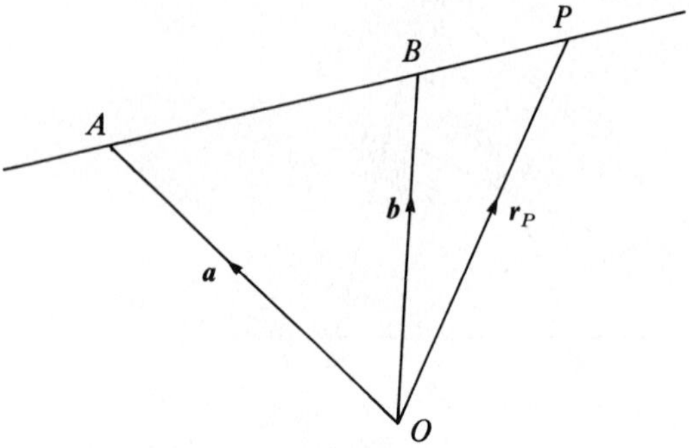

$$\boldsymbol{r}_P = \boldsymbol{a} + t(\boldsymbol{b} - \boldsymbol{a}) \quad (t \in \mathbb{R}).$$

(ii) Given one point A with position vector $\boldsymbol{a}$, and a vector $\boldsymbol{v}$ in the direction of the line.

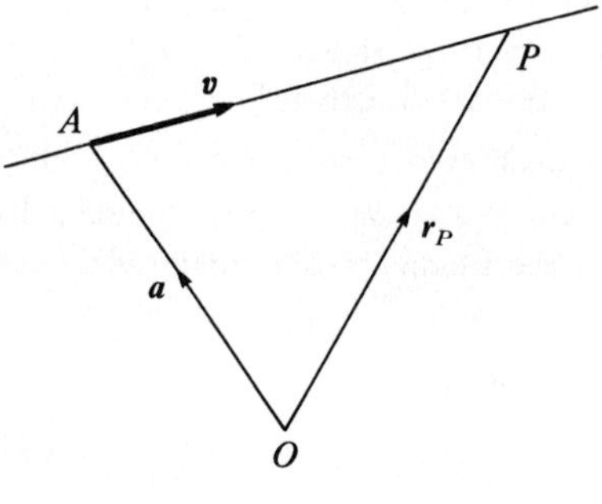

$$\boldsymbol{r}_P = \boldsymbol{a} + t\boldsymbol{v} \quad (t \in \mathbb{R}).$$

10

Straight lines and planes

In three dimensions a straight line may be specified by either

(i) two distinct points on it, or

(ii) a point on it and its direction.

These give rise to the vector form of equation for a line as follows. For diagrams, see Example 10.1. First, given points A and B, with position vectors $\boldsymbol{a}$ and $\boldsymbol{b}$, let P be any point on the straight line. The Triangle Law gives

$$\boldsymbol{r}_P = \overrightarrow{OA} + \overrightarrow{AP}.$$

Now $\overrightarrow{AP}$ is in the same direction as (or in the opposite direction to) $\overrightarrow{AB}$. Hence $\overrightarrow{AP}$ is a multiple of $\overrightarrow{AB}$, say $\overrightarrow{AP} = t\,\overrightarrow{AB}$. We know that $\overrightarrow{OA} = \boldsymbol{a}$ and $\overrightarrow{AB} = \boldsymbol{b} - \boldsymbol{a}$, so

$$\boldsymbol{r}_P = \boldsymbol{a} + t(\boldsymbol{b} - \boldsymbol{a}). \qquad (1)$$

Second, given a point A with position vector $\boldsymbol{a}$, and a direction, say the direction specified by the vector $\boldsymbol{v}$, let P be any point on the straight line. Then

$$\boldsymbol{r}_P = \overrightarrow{OA} + \overrightarrow{AP}, \quad \text{as above,}$$

but here $\overrightarrow{AP}$ is a multiple of $\boldsymbol{v}$, say $\overrightarrow{AP} = t\boldsymbol{v}$. Hence

$$\boldsymbol{r}_P = \boldsymbol{a} + t\boldsymbol{v}. \qquad (2)$$

10.2 To find parametric equations for given straight lines.
(i) Line through $A(2,3,-1)$ and $B(3,1,3)$.

$$\boldsymbol{a}=\begin{bmatrix}2\\3\\-1\end{bmatrix},\quad \boldsymbol{b}=\begin{bmatrix}3\\1\\3\end{bmatrix},\quad \text{so}\quad \boldsymbol{b}-\boldsymbol{a}=\begin{bmatrix}1\\-2\\4\end{bmatrix}.$$

An equation for the line is

$$\begin{bmatrix}x\\y\\z\end{bmatrix}=\begin{bmatrix}2\\3\\-1\end{bmatrix}+t\begin{bmatrix}1\\-2\\4\end{bmatrix}\quad (t\in\mathbb{R}).$$

In terms of coordinates, this becomes

$$x=2+t,\quad y=3-2t,\quad z=-1+4t\quad (t\in\mathbb{R}).$$

(ii) Line through $A(-2,5,1)$ in the direction of the vector

$$\boldsymbol{v}=\begin{bmatrix}1\\-1\\2\end{bmatrix}.$$

An equation for the line is

$$\begin{bmatrix}x\\y\\z\end{bmatrix}=\begin{bmatrix}-2\\5\\1\end{bmatrix}+t\begin{bmatrix}1\\-1\\2\end{bmatrix}\quad (t\in\mathbb{R}).$$

In terms of coordinates, this becomes

$$x=-2+t,\quad y=5-t,\quad z=1+2t\quad (t\in\mathbb{R}).$$

10.3 Find parametric equations for the straight line through $A(-2,5,1)$ in the direction of the unit vector

$$\boldsymbol{u}=\begin{bmatrix}\frac{1}{\sqrt{6}}\\-\frac{1}{\sqrt{6}}\\\frac{2}{\sqrt{6}}\end{bmatrix}.$$

An equation for the line is

$$\begin{bmatrix}x\\y\\z\end{bmatrix}=\begin{bmatrix}-2\\5\\1\end{bmatrix}+t\begin{bmatrix}\frac{1}{\sqrt{6}}\\-\frac{1}{\sqrt{6}}\\\frac{2}{\sqrt{6}}\end{bmatrix}\quad (t\in\mathbb{R}).$$

This of course is the *same line* as in Example 10.2 (ii), because the vector $\boldsymbol{u}$ is in the same direction as the vector $\boldsymbol{v}$ given there. The equations obtained look different, but as t varies in each case the sets of equations determine the same sets of values for x, y and z. For example, the point $(-1,4,3)$ arises from the equations in Example 10.2 (ii) with $t=1$, and from the equations above with $t=\sqrt{6}$.

Equations (1) and (2) are forms of vector equation for a straight line. The number t is a parameter: as t varies, the right-hand side gives the position vectors of the points on the line. The equation (1) is in fact a special case of (2). We may represent the vector $\boldsymbol{r}_P$ by

$$\begin{bmatrix} x \\ y \\ z \end{bmatrix},$$

where (x, y, z) are the coordinates of P, and then if

$$\boldsymbol{v} = \begin{bmatrix} v_1 \\ v_2 \\ v_3 \end{bmatrix},$$

(2) becomes

$$\begin{bmatrix} x \\ y \\ z \end{bmatrix} = \begin{bmatrix} a_1 \\ a_2 \\ a_3 \end{bmatrix} + t \begin{bmatrix} v_1 \\ v_2 \\ v_3 \end{bmatrix}$$

(say), or

$$\left.\begin{aligned} x &= a_1 + tv_1 \\ y &= a_2 + tv_2 \\ z &= a_3 + tv_3 \end{aligned}\right\} \quad (t \in \mathbb{R}),$$

which is the coordinate form of parametric equations for a straight line. Examples 10.2 give specific cases.

The components of a vector such as $\boldsymbol{v}$ above, when used to specify a direction, are called *direction ratios*. In this situation *any* vector in the particular direction will serve the purpose. It is often convenient, however, to use a unit vector. See Example 10.3.

10.4 To find whether straight lines with given parametric equations intersect.

Line 1: $x=2-t, \quad y=3t, \quad z=2+2t \quad (t \in \mathbb{R})$.

Line 2: $x=-3-4u, \quad y=4+u, \quad z=1-3u \quad (u \in \mathbb{R})$.

At a point of intersection we would have a value of t and a value of u satisfying:

$$\left.\begin{aligned} 2-t&=-3-4u \\ 3t&=4+u \\ 2+2t&=1-3u \end{aligned}\right\}. \quad \text{i.e.} \quad \left.\begin{aligned} t-4u&=5 \\ 3t-u&=4 \\ 2t+3u&=-1 \end{aligned}\right\}.$$

Here we have three equations in two unknowns. We may expect them to be inconsistent, in which case the two lines have no point of intersection. Let us find out, using the GE process.

$$\begin{bmatrix} 1 & -4 & 5 \\ 3 & -1 & 4 \\ 2 & 3 & -1 \end{bmatrix} \to \begin{bmatrix} 1 & -4 & 5 \\ 0 & 11 & -11 \\ 0 & 11 & -11 \end{bmatrix} \to \begin{bmatrix} 1 & -4 & 5 \\ 0 & 1 & -1 \\ 0 & 0 & 0 \end{bmatrix}.$$

So the equations are *consistent*, and there is a unique solution $u=-1$, $t=1$. Substitute either of these into the equations for the appropriate line, to obtain

$$x=1, \quad y=3, \quad z=4,$$

the coordinates of the intersection point.

10.5 A proof that for any non-zero vectors $\boldsymbol{a}, \boldsymbol{b}$,

$$\boldsymbol{a}\,.\,\boldsymbol{b}=|\boldsymbol{a}||\boldsymbol{b}|\cos\theta,$$

where θ is the angle between $\boldsymbol{a}$ and $\boldsymbol{b}$. Let

$$\boldsymbol{a}=\begin{bmatrix} a_1 \\ a_2 \\ a_3 \end{bmatrix}, \quad \text{and} \quad \boldsymbol{b}=\begin{bmatrix} b_1 \\ b_2 \\ b_3 \end{bmatrix},$$

and let A and B be points with position vectors $\boldsymbol{a}$ and $\boldsymbol{b}$ respectively.

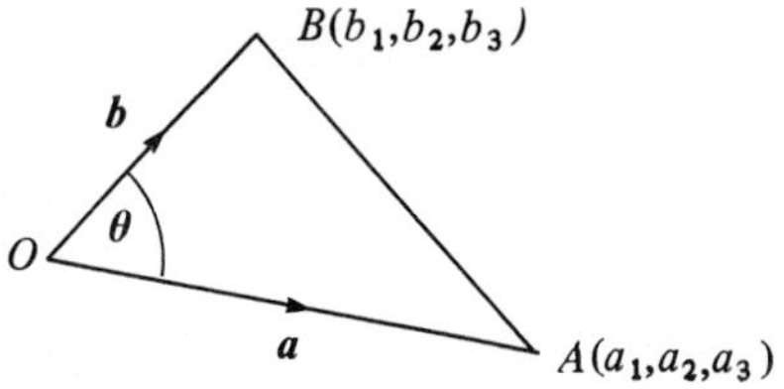

Then $A\hat{O}B=\theta$, and in triangle OAB the cosine formula gives

$$\begin{aligned} |AB|^2&=|OA|^2+|OB|^2-2|OA||OB|\cos\theta \\ &=|\boldsymbol{a}|^2+|\boldsymbol{b}|^2-2|\boldsymbol{a}||\boldsymbol{b}|\cos\theta. \end{aligned}$$

Hence

$$\begin{aligned} |\boldsymbol{a}||\boldsymbol{b}|\cos\theta&=\tfrac{1}{2}(|\boldsymbol{a}|^2+|\boldsymbol{b}|^2-|AB|^2) \\ &=\tfrac{1}{2}(a_1{}^2+a_2{}^2+a_3{}^2+b_1{}^2+b_2{}^2+b_3{}^2 \\ &\qquad -(b_1-a_1)^2-(b_2-a_2)^2-(b_3-a_3)^2) \\ &=\tfrac{1}{2}(2a_1b_1+2a_2b_2+2a_3b_3) \\ &=a_1b_1+a_2b_2+a_3b_3 \\ &=\boldsymbol{a}\,.\,\boldsymbol{b}. \end{aligned}$$

Two straight lines in three dimensions need not intersect. Example 10.4 shows how, given vector equations for two lines, we can check whether they intersect and, if they do, how to find the point of intersection. Note that, to avoid confusion, we use different letters for the parameters in the equations for the two lines.

There is a way of dealing with angles in three dimensions, using vectors. This involves the idea of the dot product of two vectors. By the *angle between two vectors* $\boldsymbol{a}$ and $\boldsymbol{b}$ we mean the angle $A\hat{O}B$, where $\overrightarrow{OA}=\boldsymbol{a}$ and $\overrightarrow{OB}=\boldsymbol{b}$. (Note that this angle is the same as any angle $A\hat{P}B$ where $\overrightarrow{PA}=\boldsymbol{a}$ and $\overrightarrow{PB}=\boldsymbol{b}$.) This angle always lies between 0 and π radians (180°) inclusive. In algebraic terms, the *dot product* of two 3-vectors

$$\boldsymbol{a}=\begin{bmatrix} a_1 \\ a_2 \\ a_3 \end{bmatrix} \quad \text{and} \quad \boldsymbol{b}=\begin{bmatrix} b_1 \\ b_2 \\ b_3 \end{bmatrix}$$

is defined by

$$\boldsymbol{a}\,.\,\boldsymbol{b}=a_1b_1+a_2b_2+a_3b_3.$$

Notice that the value of a dot product is a number, not a vector. In some books the dot product is referred to as the *scalar* product. What has this to do with angles?

Rule

Let $\boldsymbol{a}$ and $\boldsymbol{b}$ be non-zero vectors and let θ be the angle between them. Then

$$\boldsymbol{a}\,.\,\boldsymbol{b}=|\boldsymbol{a}|\,|\boldsymbol{b}|\cos\theta.$$

Consequently,

$$\cos\theta=\frac{\boldsymbol{a}\,.\,\boldsymbol{b}}{|\boldsymbol{a}||\boldsymbol{b}|}.$$

A demonstration of this rule is given as Example 10.5.

10.6 Calculate the cosine of the angle θ between the given vectors in each case.

(i)
$$\boldsymbol{a}=\begin{bmatrix}1\\1\\2\end{bmatrix},\quad \boldsymbol{b}=\begin{bmatrix}-1\\2\\1\end{bmatrix}.$$

$$|\boldsymbol{a}|=\sqrt{6},\quad |\boldsymbol{b}|=\sqrt{6}.$$
$$\boldsymbol{a}.\boldsymbol{b}=-1+2+2=3.$$

Hence

$$\cos\theta=\frac{3}{\sqrt{6}\times\sqrt{6}}=\frac{1}{2}.$$

(ii)
$$\boldsymbol{a}=\begin{bmatrix}-2\\3\\1\end{bmatrix},\quad \boldsymbol{b}=\begin{bmatrix}1\\2\\2\end{bmatrix}.$$

$$|\boldsymbol{a}|=\sqrt{14},\quad |\boldsymbol{b}|=3.$$
$$\boldsymbol{a}.\boldsymbol{b}=-2+6+2=6.$$

Hence

$$\cos\theta=\frac{6}{\sqrt{14}\times 3}=\frac{2}{\sqrt{14}}.$$

(iii)
$$\boldsymbol{a}=\begin{bmatrix}-1\\1\\-4\end{bmatrix},\quad \boldsymbol{b}=\begin{bmatrix}2\\3\\1\end{bmatrix}.$$

$$|\boldsymbol{a}|=\sqrt{18},\quad |\boldsymbol{b}|=\sqrt{14}.$$
$$\boldsymbol{a}.\boldsymbol{b}=-2+3-4=-3.$$

Hence

$$\cos\theta=\frac{-3}{\sqrt{18}\times\sqrt{14}}=-\frac{1}{2\times\sqrt{7}}.$$

(The negative sign indicates an *obtuse* angle.)

10.7 Proof that, for any three vectors $\boldsymbol{a}$, $\boldsymbol{b}$ and $\boldsymbol{c}$,

$$\boldsymbol{a}.(\boldsymbol{b}+\boldsymbol{c})=\boldsymbol{a}.\boldsymbol{b}+\boldsymbol{a}.\boldsymbol{c}.$$

Let

$$\boldsymbol{a}=\begin{bmatrix}a_1\\a_2\\a_3\end{bmatrix},\quad \boldsymbol{b}=\begin{bmatrix}b_1\\b_2\\b_3\end{bmatrix},\quad \boldsymbol{c}=\begin{bmatrix}c_1\\c_2\\c_3\end{bmatrix}.$$

Then

$$\boldsymbol{b}+\boldsymbol{c}=\begin{bmatrix}b_1+c_1\\b_2+c_2\\b_3+c_3\end{bmatrix}.$$

So

$$\begin{aligned}\boldsymbol{a}.(\boldsymbol{b}+\boldsymbol{c})&=a_1(b_1+c_1)+a_2(b_2+c_2)+a_3(b_3+c_3)\\&=a_1b_1+a_1c_1+a_2b_2+a_2c_2+a_3b_3+a_3c_3\\&=a_1b_1+a_2b_2+a_3b_3+a_1c_1+a_2c_2+a_3c_3\\&=\boldsymbol{a}.\boldsymbol{b}+\boldsymbol{a}.\boldsymbol{c}.\end{aligned}$$

Using this rule, we are now able to calculate (the cosines of) angles between given vectors. See Examples 10.6. One of these examples illustrates a general rule. A dot product $\boldsymbol{a}\,.\,\boldsymbol{b}$ may equal zero even though neither $\boldsymbol{a}$ nor $\boldsymbol{b}$ is itself $\mathbf{0}$.

Rule

(i) If the angle between two non-zero vectors $\boldsymbol{a}$ and $\boldsymbol{b}$ is a right angle, then $\boldsymbol{a}\,.\,\boldsymbol{b}=0$.
(ii) If $\boldsymbol{a}$ and $\boldsymbol{b}$ are non-zero vectors with $\boldsymbol{a}\,.\,\boldsymbol{b}=0$, the angle between $\boldsymbol{a}$ and $\boldsymbol{b}$ is a right angle.

Two non-zero vectors are said to be *perpendicular* (or *orthogonal*) if the angle between them is a right angle.

Rules

For dot products:

(i) $\boldsymbol{a}\,.\,\boldsymbol{b}=\boldsymbol{b}\,.\,\boldsymbol{a}$ for all vectors $\boldsymbol{a}$ and $\boldsymbol{b}$.
(ii) $\boldsymbol{a}\,.\,\boldsymbol{a}=|\boldsymbol{a}|^2$ for all vectors $\boldsymbol{a}$.
(iii) $\boldsymbol{a}\,.\,(\boldsymbol{b}+\boldsymbol{c})=\boldsymbol{a}\,.\,\boldsymbol{b}+\boldsymbol{a}\,.\,\boldsymbol{c}$ for all vectors $\boldsymbol{a}, \boldsymbol{b}, \boldsymbol{c}$.
(iv) $(k\boldsymbol{a})\,.\,\boldsymbol{b}=\boldsymbol{a}\,.\,(k\boldsymbol{b})=k(\boldsymbol{a}\,.\,\boldsymbol{b})$ for all vectors $\boldsymbol{a}, \boldsymbol{b}, \boldsymbol{c}$, and all $k\in\mathbb{R}$.
(v) $\boldsymbol{i}\,.\,\boldsymbol{j}=\boldsymbol{j}\,.\,\boldsymbol{k}=\boldsymbol{k}\,.\,\boldsymbol{i}=0$.

These are quite straightforward. Part (iii) is proved in Example 10.7. The others may be regarded as exercises.

10.8 A plane through a given point A, perpendicular to a given vector $\boldsymbol{n}$.

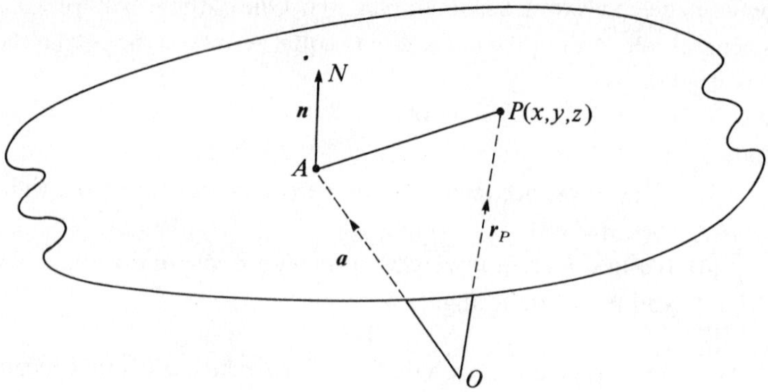

Let A be the point $(3, 1, -2)$, and let $\boldsymbol{n}$ be the vector

$$\begin{bmatrix} -1 \\ 2 \\ -1 \end{bmatrix}.$$

The point $P(x, y, z)$ lies on the plane if and only if AP is perpendicular to AN, i.e. if and only if $\overrightarrow{AP} \cdot \boldsymbol{n}=0$. Now

$$\overrightarrow{AP}=\overrightarrow{OP}-\overrightarrow{OA}=\begin{bmatrix} x \\ y \\ z \end{bmatrix}-\begin{bmatrix} 3 \\ 1 \\ -2 \end{bmatrix}=\begin{bmatrix} x-3 \\ y-1 \\ z+2 \end{bmatrix}.$$

So $\overrightarrow{AP} \cdot \boldsymbol{n}=0$ becomes

$$(x-3)(-1)+(y-1)2+(z+2)(-1)=0.$$

i.e. $$-x+3+2y-2-z-2=0.$$

i.e. $$-x+2y-z-1=0.$$

This is an equation for the plane in this case.

10.9 Show that the vector $\boldsymbol{n}=\begin{bmatrix} a \\ b \\ c \end{bmatrix}$ is perpendicular to the plane with equation $ax+by+cz+d=0$.

Let $P_1(x_1, y_1, z_1)$ and $P_2(x_2, y_2, z_2)$ be two points lying on the plane. Then

$$\left.\begin{aligned} &ax_1+by_1+cz_1+d=0 \\ \text{and} \quad &ax_2+by_2+cz_2+d=0. \end{aligned}\right\} \quad (*)$$

We show that P_1P_2 is perpendicular to $\boldsymbol{n}$.

$$\begin{aligned} \boldsymbol{n} \cdot \overrightarrow{P_1P_2} &= a(x_2-x_1)+b(y_2-y_1)+c(z_2-z_1) \\ &= ax_2-ax_1+by_2-by_1+cz_2-cz_1 \\ &= ax_2+by_2+cz_2-(ax_1+by_1+cz_1) \\ &= -d-(-d) \quad \text{by (*) above} \\ &= 0. \end{aligned}$$

Hence $\boldsymbol{n}$ is perpendicular to P_1P_2 and, since P_1 and P_2 were chosen arbitrarily, $\boldsymbol{n}$ is perpendicular to the plane.

A *plane* in three dimensions may be specified by either

(i) three points on the plane, or

(ii) one point on the plane and a vector in the direction perpendicular to the plane.

We postpone consideration of (i) until later (see Example 11.9). The procedure for (ii) is as follows. A diagram is given in Example 10.8.

Let A be the point (a_1, a_2, a_3) and let $\boldsymbol{n}$ be a vector (which is to specify the direction perpendicular to the plane). Let $\overrightarrow{OA}=\boldsymbol{a}$, and let N be such that $\overrightarrow{AN}=\boldsymbol{n}$. The point $P(x, y, z)$ lies on the plane through A perpendicular to AN if and only if AP is perpendicular to AN, i.e. vector $\overrightarrow{AP}$ is perpendicular to vector $\overrightarrow{AN}$,

i.e. $$\overrightarrow{AP} \cdot \boldsymbol{n}=0,$$

i.e. $$(\overrightarrow{OP}-\overrightarrow{OA}) \cdot \boldsymbol{n}=0,$$

i.e. $$(\boldsymbol{r}_P-\boldsymbol{a}) \cdot \boldsymbol{n}=0.$$

This last is a vector form of equation for the plane (not a parametric equation this time, though). The vector $\boldsymbol{n}$ is called a *normal* to the plane.

The equation which we have just derived can be written in coordinate form. Let

$$\boldsymbol{n}=\begin{bmatrix} n_1 \\ n_2 \\ n_3 \end{bmatrix}.$$

Then the equation becomes

$$(x-a_1)n_1+(y-a_2)n_2+(z-a_3)n_3=0.$$

Example 10.8 contains a specific case of this.

Rule

Equations of planes have the form

$$ax+by+cz+d=0.$$

Example 10.9 shows that, given such an equation, we can read off a normal vector, namely

$$\begin{bmatrix} a \\ b \\ c \end{bmatrix}.$$

10.10 The angle between two planes.

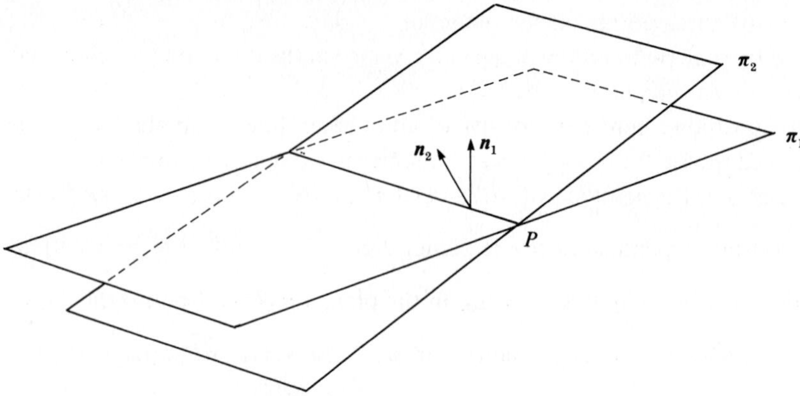

Plane π_1 has normal vector $\boldsymbol{n}_1$.
Plane π_2 has normal vector $\boldsymbol{n}_2$.
Here is an 'edge-on' view, looking along the line of intersection.

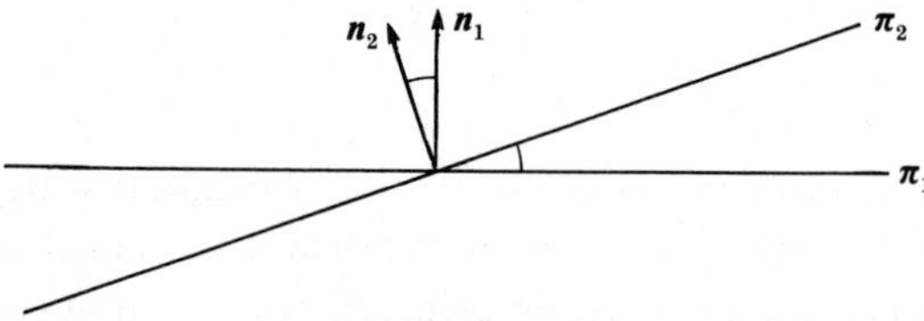

The angle observed here between the planes is the same (by a simple geometrical argument) as the angle between the normal vectors.

Let π_1 and π_2 have equations respectively

$$x - y + 3z + 2 = 0$$

and

$$-x + 2y + 2z - 3 = 0.$$

The two normal vectors can then be taken as

$$\boldsymbol{n}_1 = \begin{bmatrix} 1 \\ -1 \\ 3 \end{bmatrix} \quad \text{and} \quad \boldsymbol{n}_2 = \begin{bmatrix} -1 \\ 2 \\ 2 \end{bmatrix},$$

say. The angle between these vectors is θ, where

$$\cos\theta = \frac{\boldsymbol{n}_1 \cdot \boldsymbol{n}_2}{|\boldsymbol{n}_1||\boldsymbol{n}_2|} = \frac{-1-2+6}{\sqrt{11} \times \sqrt{3}} = \frac{3}{\sqrt{33}}.$$

10.11 Find parametric equations for the line of intersection of the two planes with equations:

$$x - y + 3z + 2 = 0,$$
$$-x + 2y + 2z - 3 = 0.$$

Try to solve the equations

Two planes will normally intersect in a straight line. The *angle* between two planes is defined to be the angle between their normal vectors. See Example 10.10.

Planes which do not intersect are parallel. We can tell immediately from their equations whether two given planes are parallel, since parallel planes have normal vectors in the same direction.

$$3x+\ y-2z+1=0$$

and

$$6x+2y-4z+5=0$$

are equations of parallel planes. By inspection we can see that they have normal vectors

$$\begin{bmatrix} 3 \\ 1 \\ -2 \end{bmatrix} \quad \text{and} \quad \begin{bmatrix} 6 \\ 2 \\ -4 \end{bmatrix}$$

respectively. These vectors are clearly in the same direction, being multiples of one another. But we can also see quite easily that the two equations above are inconsistent. If we tried to solve them simultaneously we would find that there are no solutions. Of course this is to be expected: any solutions would yield coordinates for points common to the two planes.

Distinct planes which do intersect have a line in common. Parametric equations for the line of intersection are given by the standard process for solving sets of equations. We would expect a set of two equations with three unknowns (if it had solutions at all) to have infinitely many solutions, these being specified by expressions involving a single parameter. An example is given in Example 10.11.

Here is a clear situation where algebra and geometry impinge. And it becomes clearer when we consider the ways in which *three* planes might intersect. A point which is common to three planes has coordinates which satisfy three linear equations simultaneously. So finding the intersection of three planes amounts to solving a set of three linear equations in three unknowns. As we know, there can be three possible outcomes.

(i) The equations may be inconsistent. In this case the planes have no common point. Either they are all parallel or each is parallel to the line of intersection of the other two.

(ii) The equations may have a unique solution. This is the general case from the geometrical point of view. The line of intersection of two of the planes meets the third plane in a single point, common to all three planes.

(iii) The equations may have infinitely many solutions, which may be specified using one parameter or two parameters. In the first case the planes have a line in common, and in the second case the

$$x - y + 3z = -2,$$
$$-x + 2y + 2z = 3$$

simultaneously. The GE process leads to

$$\begin{bmatrix} 1 & -1 & 3 & -2 \\ 0 & 1 & 5 & 1 \end{bmatrix},$$

and hence to a solution

$$z = t, \quad y = 1 - 5t, \quad x = -1 - 8t \quad (t \in \mathbb{R}).$$

These are in fact just parametric equations for the line of intersection, as required.

10.12 Ways in which three planes can intersect.

(i) Three planes:

$$x + 2y + 3z = 0,$$
$$3x + y - z = 5,$$
$$x - y + z = 2.$$

The GE process leads to

$$\begin{bmatrix} 1 & 2 & 3 & 0 \\ 0 & 1 & 2 & -1 \\ 0 & 0 & 0 & 1 \end{bmatrix}.$$

Hence the set of equations is inconsistent. There is no point common to the three planes. Nevertheless each pair of these planes has a line of intersection. The three lines of intersection are parallel. What makes this so? If you are not sure, work out their parametric equations and confirm that they all have the same direction.

(ii) Three planes:

$$x - y - z = 4,$$
$$2x - 3y + 4z = -5,$$
$$-x + 2y - 2z = 3.$$

The GE process leads to

$$\begin{bmatrix} 1 & -1 & -1 & 4 \\ 0 & 1 & -6 & 13 \\ 0 & 0 & 1 & -2 \end{bmatrix}.$$

Thus there is a unique solution $x = 3, y = 1, z = -2$. These planes have a single point in common, namely $(3, 1, -2)$. Each pair of planes intersects in a line which meets the third plane at this point.

(iii) Three planes:

$$x + 2y + z = 6,$$
$$-x + y - 4z = 3,$$
$$x - 3y + 6z = -9.$$

The GE process leads to

$$\begin{bmatrix} 1 & 2 & 1 & 6 \\ 0 & 1 & -1 & 3 \\ 0 & 0 & 0 & 0 \end{bmatrix}.$$

Hence there are infinitely many solutions, which can be specified by parametric equations

$$x = -3t, \quad y = 3 + t, \quad z = t \quad (t \in \mathbb{R}).$$

The straight line with these parametric equations is common to all three planes.

planes are all the same plane. Illustrations are provided in Example 10.12.

There is a convenient formula for finding the perpendicular distance from a given point to a given plane. Let P be the point with coordinates (x_0, y_0, z_0), and let

$$ax+by+cz+d=0$$

be an equation of a plane. Not all of a, b and c can be zero, or else this equation would not be sensible.

Rule

The perpendicular distance from the point P to the plane given above is equal to

$$\frac{|ax_0+by_0+cz_0+d|}{\sqrt{a^2+b^2+c^2}}.$$

This formula can be derived using the methods of this chapter, and this is done in Example 10.13.

Summary

A vector form of equation for a straight line in three dimensions is derived and used in geometric deductions. The dot product of two vectors is defined, and properties of it are derived. A standard form of equation for a plane is established, and ideas of linear algebra are used in considering the nature of the intersection of two or three planes. Angles between lines and between planes are dealt with. A formula is given for the perpendicular distance from a point to a plane.

Exercises

1. Find parametric equations for the straight line passing through the points A and B in each case.
   (i) $A(0,1,3)$, $B(1,0,1)$.
   (ii) $A(1,1,-2)$, $B(1,2,0)$.
   (iii) $A(-1,2,4)$, $B(-1,2,-7)$.
   (iv) $A(1,1,1)$, $B(2,2,2)$.
   (v) $A(0,0,0)$, $B(3,-1,2)$.
2. In each case below, write down a vector in the direction of the straight line with the given parametric equations.
   (i) $x=3-t$, $y=-1+2t$, $z=4-5t$ $(t\in\mathbb{R})$.
   (ii) $x=2t$, $y=1-t$, $z=2+t$ $(t\in\mathbb{R})$.
   (iii) $x=1-3t$, $y=2$, $z=3-t$ $(t\in\mathbb{R})$.

10.13 Proof that the perpendicular distance from the point $P(x_0, y_0, z_0)$ to the plane with equation $ax+by+cz+d=0$ is

$$\frac{|ax_0+by_0+cz_0+d|}{\sqrt{a^2+b^2+c^2}}.$$

The vector

$$\begin{bmatrix} a \\ b \\ c \end{bmatrix}$$

is perpendicular to the plane, so the straight line through P perpendicular to the plane has parametric equations

$$x=x_0+at, \quad y=y_0+bt, \quad z=z_0+ct \quad (t\in\mathbb{R}).$$

This line meets the plane at the point M (say), whose coordinates are given by these parametric equations with the value of t given by

$$a(x_0+at)+b(y_0+bt)+c(z_0+ct)+d=0.$$

Solve for t, obtaining

$$(a^2+b^2+c^2)t=-ax_0-by_0-cz_0-d,$$

so

$$t=-\frac{ax_0+by_0+cz_0+d}{a^2+b^2+c^2}.$$

Now

$$\begin{aligned} |PM|^2 &= (x_0+at-x_0)^2+(y_0+bt-y_0)^2+(z_0+ct-z_0)^2 \\ &= a^2t^2+b^2t^2+c^2t^2 \\ &= (a^2+b^2+c^2)t^2 \\ &= (a^2+b^2+c^2)\frac{(ax_0+by_0+cz_0+d)^2}{(a^2+b^2+c^2)^2}. \end{aligned}$$

Hence

$$|PM|=\frac{|ax_0+by_0+cz_0+d|}{\sqrt{a^2+b^2+c^2}},$$

as required.

3. In each case below, find whether the two lines with the given parametric equations intersect and, if they do, find the coordinates of the points of intersection.

   (i) $x=2+2t,\quad y=2+t,\quad z=-t \quad (t\in\mathbb{R}),$
   $x=-2+3u,\quad y=-3+6u,\quad z=7-9u \quad (u\in\mathbb{R}).$

   (ii) $x=1+t,\quad y=2-t,\quad z=-1-2t \quad (t\in\mathbb{R}),$
   $x=1+2u,\quad y=-6u,\quad z=1 \quad (u\in\mathbb{R}).$

   (iii) $x=2-t,\quad y=-1-3t,\quad z=2+2t \quad (t\in\mathbb{R}),$
   $x=u,\quad y=-2+u,\quad z=1-u \quad (u\in\mathbb{R}).$

   (iv) $x=2+t,\quad y=-t,\quad z=1+2t \quad (t\in\mathbb{R}),$
   $x=4-2u,\quad y=-2+2u,\quad z=5-4u \quad (u\in\mathbb{R}).$

4. Calculate the cosine of the angle $A\hat{P}B$ in each case, where A, P and B are the points given.

   (i) $A(1,1,1),\quad P(0,0,0),\quad B(1,1,0).$

   (ii) $A(-2,1,-1),\quad P(0,0,0),\quad B(1,2,1).$

   (iii) $A(3,-1,2),\quad P(1,2,3),\quad B(0,1,-2).$

   (iv) $A(6,7,8),\quad P(0,1,2),\quad B(0,0,1).$

5. Find the cosines of the internal angles of the triangle whose vertices are $A(1,3,-1)$, $B(0,2,-1)$, and $C(2,5,1)$, and find the radian measure of the largest of these angles.

6. Find a vector perpendicular to both $\boldsymbol{a}$ and $\boldsymbol{b}$, where

$$\boldsymbol{a}=\begin{bmatrix}2\\-1\\0\end{bmatrix} \quad \text{and} \quad \boldsymbol{b}=\begin{bmatrix}1\\1\\-1\end{bmatrix}.$$

7. Find the length of the vector $2\boldsymbol{a}+\boldsymbol{b}$, where $\boldsymbol{a}$ and $\boldsymbol{b}$ are unit vectors and the angle between them is $\pi/3$ (i.e. 60°).

8. In each case below, find the length of the perpendicular from the given point to the straight line with the given parametric equations.

   (i) $(2,1,-1),\quad x=3-t,\quad y=1+2t,\quad z=t \quad (t\in\mathbb{R}).$

   (ii) $(0,1,4),\quad x=2t,\quad y=3,\quad z=4-2t \quad (t\in\mathbb{R}).$

   (iii) $(2,3,1),\quad x=2t,\quad y=3t,\quad z=2-t \quad (t\in\mathbb{R}).$

9. In each case below, find an equation for the plane through the given point A, with normal in the direction of the given vector $\boldsymbol{n}$.

   (i) $A(3,2,1),\quad \boldsymbol{n}=\begin{bmatrix}-1\\1\\-2\end{bmatrix}.$ (ii) $A(0,1,-1),\quad \boldsymbol{n}=\begin{bmatrix}4\\5\\6\end{bmatrix}.$

   (iii) $A(-2,3,5),\quad \boldsymbol{n}=\begin{bmatrix}0\\1\\3\end{bmatrix}.$ (iv) $A(1,1,1),\quad \boldsymbol{n}=\begin{bmatrix}1\\1\\1\end{bmatrix}.$

10. Find the cosine of the acute angle between the planes with given equations in each case.

(i) $x+y+z-3=0,$
$2x-y-z+1=0.$

(ii) $3x+y+2=0,$
$x+y-4z-3=0.$

(iii) $x+y=0,$
$y+z=0.$

(iv) $3x-y+2z-7=0,$
$z=0.$

11. In each case below, determine whether the intersection of the three planes with given equations is empty, is a single point, or is a straight line.

(i) $x-2y-3z=-1,$
$y+2z=1,$
$2x+y+4z=3.$

(ii) $x+3z=-1,$
$2x-y+z=2$
$x+2y-z=5.$

(iii) $x+3y+5z=0,$
$x+2y+3z=1,$
$x-z=-1.$

(iv) $x-2y+z=-6,$
$-2+z=8,$
$x+2y+2z=1.$

12. In each case below, find the perpendicular distance from the given point P to the plane with the given equation.

(i) $P(2,1,2),\quad x-2y-z+2=0.$

(ii) $P(-1,0,1),\quad 3x-y+2z-5=0.$

(iii) $P(1,1,1),\quad x+y+z=0.$

(iv) $P(0,0,0),\quad 5x-y-4z+6=0.$

(v) $P(0,0,0),\quad 5x-y-4z+1=0.$

Examples

11.1 Evaluation of cross products.

(i) $$\boldsymbol{a}=\begin{bmatrix}1\\2\\3\end{bmatrix},\quad \boldsymbol{b}=\begin{bmatrix}2\\3\\5\end{bmatrix},\quad \boldsymbol{a}\times\boldsymbol{b}=\begin{bmatrix}10-9\\6-5\\3-4\end{bmatrix}=\begin{bmatrix}1\\1\\-1\end{bmatrix}.$$

(ii) $$\boldsymbol{a}=\begin{bmatrix}2\\1\\0\end{bmatrix},\quad \boldsymbol{b}=\begin{bmatrix}-1\\3\\2\end{bmatrix},\quad \boldsymbol{a}\times\boldsymbol{b}=\begin{bmatrix}2-0\\0-4\\6+1\end{bmatrix}=\begin{bmatrix}2\\-4\\7\end{bmatrix}.$$

(iii) $$\boldsymbol{a}=\begin{bmatrix}2\\-1\\3\end{bmatrix},\quad \boldsymbol{b}=\begin{bmatrix}-4\\2\\-6\end{bmatrix},\quad \boldsymbol{a}\times\boldsymbol{b}=\begin{bmatrix}6-6\\-12+12\\4-4\end{bmatrix}=\begin{bmatrix}0\\0\\0\end{bmatrix}.$$

(iv) $$\boldsymbol{a}=\begin{bmatrix}-1\\3\\2\end{bmatrix},\quad \boldsymbol{b}=\begin{bmatrix}2\\1\\0\end{bmatrix},\quad \boldsymbol{a}\times\boldsymbol{b}=\begin{bmatrix}0-2\\4-0\\-1-6\end{bmatrix}=\begin{bmatrix}-2\\4\\-7\end{bmatrix}.$$

Compare (ii) with (iv).

11.2 Use of the determinant mnemonic in evaluating cross products.

(i) $\boldsymbol{a}=-\boldsymbol{i}+2\boldsymbol{j}-\boldsymbol{k},\quad \boldsymbol{b}=3\boldsymbol{i}-\boldsymbol{j}.$

$$\boldsymbol{a}\times\boldsymbol{b}=\begin{vmatrix}\boldsymbol{i}&-1&3\\\boldsymbol{j}&2&1\\\boldsymbol{k}&-1&0\end{vmatrix}=\boldsymbol{i}\begin{vmatrix}2&1\\-1&0\end{vmatrix}-\boldsymbol{j}\begin{vmatrix}-1&3\\-1&0\end{vmatrix}+\boldsymbol{k}\begin{vmatrix}-1&3\\2&1\end{vmatrix}$$
$$=\boldsymbol{i}-3\boldsymbol{j}-7\boldsymbol{k}.$$

(ii) $\boldsymbol{a}=3\boldsymbol{i}-4\boldsymbol{j}+2\boldsymbol{k},\quad \boldsymbol{b}=\boldsymbol{j}+2\boldsymbol{k}.$

$$\boldsymbol{a}\times\boldsymbol{b}=\begin{vmatrix}\boldsymbol{i}&3&0\\\boldsymbol{j}&-4&1\\\boldsymbol{k}&2&2\end{vmatrix}=\boldsymbol{i}\begin{vmatrix}-4&1\\2&2\end{vmatrix}-\boldsymbol{j}\begin{vmatrix}3&0\\2&2\end{vmatrix}+\boldsymbol{k}\begin{vmatrix}3&0\\-4&1\end{vmatrix}$$
$$=-10\boldsymbol{i}-6\boldsymbol{j}+3\boldsymbol{k}.$$

11.3 Proof that for any 3-vectors $\boldsymbol{a}$ and $\boldsymbol{b}$, $\boldsymbol{a}\times\boldsymbol{b}$ is perpendicular to $\boldsymbol{a}$ and to $\boldsymbol{b}$.

Let

$$\boldsymbol{a}=\begin{bmatrix}a_1\\a_2\\a_3\end{bmatrix}\quad\text{and}\quad \boldsymbol{b}=\begin{bmatrix}b_1\\b_2\\b_3\end{bmatrix}.$$

Then

$$\boldsymbol{a}\times\boldsymbol{b}=\begin{bmatrix}a_2b_3-a_3b_2\\a_3b_1-a_1b_3\\a_1b_2-a_2b_1\end{bmatrix},$$

so

$$\begin{aligned}\boldsymbol{a}\,.\,(\boldsymbol{a}\times\boldsymbol{b})&=a_1(a_2b_3-a_3b_2)+a_2(a_3b_1-a_1b_3)+a_3(a_1b_2-a_2b_1)\\&=a_1a_2b_3-a_1a_3b_2+a_2a_3b_1-a_2a_1b_3+a_3a_1b_2-a_3a_2b_1\\&=0.\end{aligned}$$

Similarly $\boldsymbol{b}\,.\,(\boldsymbol{a}\times\boldsymbol{b})=0$.

11

Cross product

We have dealt with the dot product of two vectors. There is another way of combining vectors which is useful in geometric applications.

Let

$$\boldsymbol{a}=\begin{bmatrix} a_1 \\ a_2 \\ a_3 \end{bmatrix} \quad \text{and} \quad \boldsymbol{b}=\begin{bmatrix} b_1 \\ b_2 \\ b_3 \end{bmatrix}$$

be two 3-vectors. The *cross product* of $\boldsymbol{a}$ and $\boldsymbol{b}$ is defined by

$$\boldsymbol{a}\times\boldsymbol{b}=\begin{bmatrix} a_2b_3-a_3b_2 \\ a_3b_1-a_1b_3 \\ a_1b_2-a_2b_1 \end{bmatrix}.$$

Note that the result is a vector this time. Example 11.1 gives some calculations. Example 11.2 shows how to apply the following mnemonic which is useful in calculating cross products. Write

$$\boldsymbol{a}=a_1\boldsymbol{i}+a_2\boldsymbol{j}+a_3\boldsymbol{k} \quad \text{and} \quad \boldsymbol{b}=b_1\boldsymbol{i}+b_2\boldsymbol{j}+b_3\boldsymbol{k},$$

where $\boldsymbol{i}$, $\boldsymbol{j}$ and $\boldsymbol{k}$ are the standard basis vectors. Then

$$\boldsymbol{a}\times\boldsymbol{b}=(a_2b_3-a_3b_2)\boldsymbol{i}+(a_3b_1-a_1b_3)\boldsymbol{j}+(a_1b_2-a_2b_1)\boldsymbol{k}.$$

This is reminiscent of the expansion of a determinant, and we can stretch the idea of a determinant to write

$$\boldsymbol{a}\times\boldsymbol{b}=\begin{vmatrix} \boldsymbol{i} & a_1 & b_1 \\ \boldsymbol{j} & a_2 & b_2 \\ \boldsymbol{k} & a_3 & b_3 \end{vmatrix}.$$

Expanding this 'determinant' by the first column gives the correct expression for $\boldsymbol{a}\times\boldsymbol{b}$.

The definition above of the cross product involved two 3-vectors explicitly. It is important to realise that, unlike the dot product, the cross product applies only to 3-vectors, and its application is of use in three-dimensional geometry.

11.4 Proof that $|\boldsymbol{a}\times\boldsymbol{b}|=|\boldsymbol{a}||\boldsymbol{b}|\sin\theta$, where $\boldsymbol{a}$ and $\boldsymbol{b}$ are non-zero 3-vectors and θ is the angle between them, for the case when $\theta\neq 0$ and $\theta\neq\pi$.

Suppose that $\boldsymbol{a}$ and $\boldsymbol{b}$ are 3-vectors, as in Example 11.3.

$$\begin{aligned}|\boldsymbol{a}\times\boldsymbol{b}|^2&=(a_2b_3-a_3b_2)^2+(a_3b_1-a_1b_3)^2+(a_1b_2-a_2b_1)^2\\&=a_2{}^2b_3{}^2+a_3{}^2b_2{}^2+a_3{}^2b_1{}^2+a_1{}^2b_3{}^2+a_1{}^2b_2{}^2+a_2{}^2b_1{}^2\\&\quad-2a_2b_3a_3b_2-2a_3b_1a_1b_3-2a_1b_2a_2b_1.\end{aligned}$$

Also

$$\begin{aligned}(|\boldsymbol{a}||\boldsymbol{b}|\sin\theta)^2&=|\boldsymbol{a}|^2|\boldsymbol{b}|^2(1-\cos^2\theta)\\&=|\boldsymbol{a}|^2|\boldsymbol{b}|^2(1-(\boldsymbol{a}\,.\,\boldsymbol{b})^2/|\boldsymbol{a}|^2|\boldsymbol{b}|^2)\\&=|\boldsymbol{a}|^2|\boldsymbol{b}|^2-(\boldsymbol{a}\,.\,\boldsymbol{b})^2\\&=(a_1{}^2+a_2{}^2+a_3{}^2)(b_1{}^2+b_2{}^2+b_3{}^2)\\&\quad-(a_1b_1+a_2b_2+a_3b_3)^2\\&=a_1{}^2b_1{}^2+a_1{}^2b_2{}^2+a_1{}^2b_3{}^2+a_2{}^2b_1{}^2+a_2{}^2b_2{}^2+a_2{}^2b_3{}^2\\&\quad+a_3{}^2b_1{}^2+a_3{}^2b_2{}^2+a_3{}^2b_3{}^2-a_1{}^2b_1{}^2-a_2{}^2b_2{}^2\\&\quad-a_3{}^2b_3{}^2-2a_1b_1a_2b_2-2a_1b_1a_3b_3-2a_2b_2a_3b_3\\&=|\boldsymbol{a}\times\boldsymbol{b}|^2.\end{aligned}$$

Now $\sin\theta$ is positive, as are $|\boldsymbol{a}|$, $|\boldsymbol{b}|$ and $|\boldsymbol{a}\times\boldsymbol{b}|$, so it follows that

$$|\boldsymbol{a}\times\boldsymbol{b}|=|\boldsymbol{a}||\boldsymbol{b}|\sin\theta.$$

11.5 Proof that $\boldsymbol{a}\times(\boldsymbol{b}+\boldsymbol{c})=(\boldsymbol{a}\times\boldsymbol{b})+(\boldsymbol{a}\times\boldsymbol{c})$, for any 3-vectors $\boldsymbol{a}$, $\boldsymbol{b}$ and $\boldsymbol{c}$. Let

$$\boldsymbol{a}=\begin{bmatrix}a_1\\a_2\\a_3\end{bmatrix},\quad \boldsymbol{b}=\begin{bmatrix}b_1\\b_2\\b_3\end{bmatrix},\quad \boldsymbol{c}=\begin{bmatrix}c_1\\c_2\\c_3\end{bmatrix}.$$

Then

$$\begin{aligned}\boldsymbol{a}\times(\boldsymbol{b}+\boldsymbol{c})&=\begin{bmatrix}a_2(b_3+c_3)-a_3(b_2+c_2)\\a_3(b_1+c_1)-a_1(b_3+c_3)\\a_1(b_2+c_2)-a_2(b_1+c_1)\end{bmatrix}\\&=\begin{bmatrix}a_2b_3-a_3b_2+a_2c_3-a_3c_2\\a_3b_1-a_1b_3+a_3c_1-a_1c_3\\a_1b_2-a_2b_1+a_1c_2-a_2c_1\end{bmatrix}\\&=(\boldsymbol{a}\times\boldsymbol{b})+(\boldsymbol{a}\times\boldsymbol{c}).\end{aligned}$$

11.6 Remember that $\boldsymbol{a}\times\boldsymbol{b}=-(\boldsymbol{b}\times\boldsymbol{a})$, for any 3-vectors $\boldsymbol{a}$ and $\boldsymbol{b}$. See Example 11.1, parts (ii) and (iv).

Here is another illustration. Let

$$\boldsymbol{a}=\begin{bmatrix}2\\-3\\1\end{bmatrix}\quad\text{and}\quad \boldsymbol{b}=\begin{bmatrix}5\\-4\\1\end{bmatrix}.$$

Then

$$\boldsymbol{a}\times\boldsymbol{b}=\begin{bmatrix}1\\3\\7\end{bmatrix}\quad\text{and}\quad \boldsymbol{b}\times\boldsymbol{a}=\begin{bmatrix}-1\\-3\\-7\end{bmatrix}=-(\boldsymbol{a}\times\boldsymbol{b}).$$

Let us now explore these geometrical uses. Some properties of the cross product will emerge as we proceed.

Rule
The product vector $\boldsymbol{a}\times\boldsymbol{b}$ is perpendicular to both $\boldsymbol{a}$ and $\boldsymbol{b}$.

To see this we just write down expanded expressions for $\boldsymbol{a}.(\boldsymbol{a}\times\boldsymbol{b})$ and $\boldsymbol{b}.(\boldsymbol{a}\times\boldsymbol{b})$. Details are in Example 11.3. Both expressions are identically zero, and the rule is therefore justified, using a result from Chapter 10.

The above rule is in fact the most useful property of the cross product, and we shall see applications of it shortly. But consider now the *length* of $\boldsymbol{a}\times\boldsymbol{b}$. It has a convenient and useful geometrical interpretation.

Rule
If $\boldsymbol{a}$ and $\boldsymbol{b}$ are non-zero vectors then

$$|\boldsymbol{a}\times\boldsymbol{b}|=|\boldsymbol{a}||\boldsymbol{b}|\sin\theta,$$

where θ is the angle between $\boldsymbol{a}$ and $\boldsymbol{b}$.

Justification of the general case (when θ is not 0 or π) is given in Example 11.4. The special case is also significant, so let us formulate it into a separate rule.

Rule
(i) $\boldsymbol{a}\times\boldsymbol{a}=\boldsymbol{0}$ for every 3-vector $\boldsymbol{a}$.
(ii) $\boldsymbol{a}\times(k\boldsymbol{a})=\boldsymbol{0}$ for every 3-vector $\boldsymbol{a}$ and any number k.

Justification of these is straightforward verification, which is left as an exercise. This rule is perhaps surprising. It suggests that the cross product behaves in ways which we might not expect. This is indeed so, and we must be careful when using it.

Rules (Properties of the cross product)
(i) $\boldsymbol{a}\times(k\boldsymbol{b})=(k\boldsymbol{a})\times\boldsymbol{b}=k(\boldsymbol{a}\times\boldsymbol{b})$, for all 3-vectors $\boldsymbol{a}$ and $\boldsymbol{b}$, and any number k.
(ii) $\boldsymbol{a}\times\boldsymbol{b}=-(\boldsymbol{b}\times\boldsymbol{a})$, for all 3-vectors $\boldsymbol{a}$ and $\boldsymbol{b}$.
(iii) $\boldsymbol{a}\times(\boldsymbol{b}+\boldsymbol{c})=(\boldsymbol{a}\times\boldsymbol{b})+(\boldsymbol{a}\times\boldsymbol{c})$, for all 3-vectors $\boldsymbol{a}$, $\boldsymbol{b}$ and $\boldsymbol{c}$.
(iv) $\boldsymbol{i}\times\boldsymbol{j}=\boldsymbol{k}$, $\boldsymbol{j}\times\boldsymbol{k}=\boldsymbol{i}$, $\boldsymbol{k}\times\boldsymbol{i}=\boldsymbol{j}$.

Demonstration of these is not difficult, using the definition. See Example 11.5 for (iii). Take note of (ii)! See Example 11.6.

Now let us see how these geometrical interpretations can be used. First we consider areas.

11.7 Areas of triangles and parallelograms.

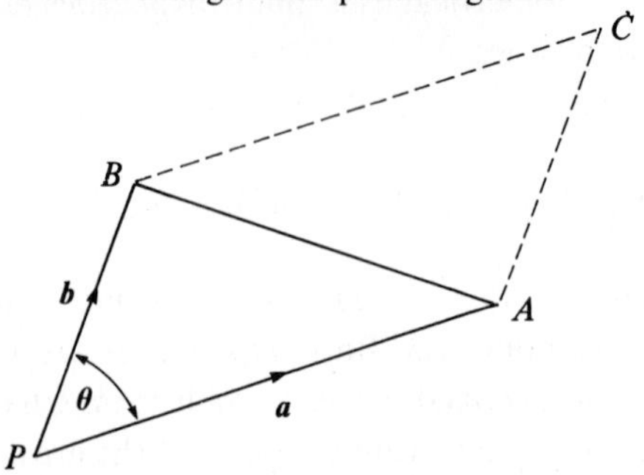

Let PAB be a triangle, with $\overrightarrow{PA}=\boldsymbol{a}$ and $\overrightarrow{PB}=\boldsymbol{b}$.
Then the area of triangle PAB is $\frac{1}{2}|\boldsymbol{a}\times\boldsymbol{b}|$.
Let C be such that $PACB$ is a parallelogram. Then the area of $PACB$ is $|\boldsymbol{a}\times\boldsymbol{b}|$.

11.8 Calculation of areas.
(i) Let

$$\boldsymbol{a}=\begin{bmatrix}2\\0\\-3\end{bmatrix},\quad \boldsymbol{b}=\begin{bmatrix}4\\1\\4\end{bmatrix}.\quad \text{Then } \boldsymbol{a}\times\boldsymbol{b}=\begin{bmatrix}3\\4\\2\end{bmatrix},$$

and $|\boldsymbol{a}\times\boldsymbol{b}|=\sqrt{9+16+4}=\sqrt{29}$.

Hence the area of the triangle PAB in the above diagram would be $\frac{1}{2}\sqrt{29}$ units^2.
(ii) Let the three points be $P(3,-1,1)$, $A(1,1,0)$ and $B(0,3,1)$. Then

$$\overrightarrow{PA}=\begin{bmatrix}-2\\2\\-1\end{bmatrix},\quad \overrightarrow{PB}=\begin{bmatrix}-3\\4\\0\end{bmatrix}\quad \text{and}\quad \overrightarrow{PA}\times\overrightarrow{PB}=\begin{bmatrix}4\\4\\-2\end{bmatrix}.$$

Hence the area of triangle PAB is equal to $\frac{1}{2}\sqrt{16+16+4}$, i.e. 3 units^2.

11.9 Find an equation for the plane through the three points $A(5,3,-1)$, $B(2,-2,0)$ and $C(3,1,1)$.

$$\overrightarrow{AB}=\begin{bmatrix}-3\\-5\\1\end{bmatrix},\quad \overrightarrow{AC}=\begin{bmatrix}-2\\-2\\2\end{bmatrix}.$$

The vector $\overrightarrow{AB}\times\overrightarrow{AC}$ is perpendicular to both $\overrightarrow{AB}$ and $\overrightarrow{AC}$, so is a normal vector to the plane of A, B and C.

$$\overrightarrow{AB}\times\overrightarrow{AC}=\begin{bmatrix}-8\\4\\-4\end{bmatrix}.$$

Any vector in the direction of this vector is a normal vector for the plane. We can choose this one or any multiple of it. Taking $\begin{bmatrix}2\\-1\\1\end{bmatrix}$, as the normal vector, we obtain an equation for the plane:

$$2(x-5)-1(y-3)+1(z+1)=0,$$

i.e. $2x-y+z-6=0$. (See Chapter 10 for the method used.)

Rule

Let $\overrightarrow{PA}$ and $\overrightarrow{PB}$ represent 3-vectors $\boldsymbol{a}$ and $\boldsymbol{b}$. Then the area of triangle PAB is equal to $\frac{1}{2}|\boldsymbol{a}\times\boldsymbol{b}|$.

This follows from our knowledge that

$$|\boldsymbol{a}\times\boldsymbol{b}|=|\boldsymbol{a}||\boldsymbol{b}|\sin\theta,$$

and the familiar rule that

$$\text{area of } \triangle PAB=\tfrac{1}{2}|PA||PB|\sin A\hat{P}B.$$

For a diagram see Example 11.7.

Further, if C is such that $PACB$ is a parallelogram, the area of the parallelogram is equal to $|\boldsymbol{a}\times\boldsymbol{b}|$. Some calculations of areas are given in Examples 11.8.

Example 11.9 gives an application of another use of cross products. A plane may be specified by giving the position vectors (or coordinates) of three points on it, say $A(\boldsymbol{a})$, $B(\boldsymbol{b})$ and $C(\boldsymbol{c})$. We know from Chapter 10 how to derive an equation for a plane given a point on it and the direction of a normal vector to it. We obtain a normal vector in the present case by using the cross product. AB represents the vector $\boldsymbol{b}-\boldsymbol{a}$ and AC represents the vector $\boldsymbol{c}-\boldsymbol{a}$. Consequently $(\boldsymbol{b}-\boldsymbol{a})\times(\boldsymbol{c}-\boldsymbol{a})$ is perpendicular to both AB and AC, and so must be perpendicular to the plane containing A, B and C. It will serve as a normal vector, and the method of Chapter 10 can now be applied.

11.10 Calculation of volumes of parallelepipeds.

(i) Find the volume of the parallelepiped with one vertex at $P(1,2,-1)$ and adjacent vertices at $A(3,-1,0)$, $B(2,1,1)$ and $C(4,0,-2)$.

The volume is $|\boldsymbol{a}.(\boldsymbol{b}\times\boldsymbol{c})|$, where $\boldsymbol{a}=\overrightarrow{PA}$, $\boldsymbol{b}=\overrightarrow{PB}$ and $\boldsymbol{c}=\overrightarrow{PC}$.

$$\overrightarrow{PA}=\begin{bmatrix}2\\-3\\1\end{bmatrix},\quad \overrightarrow{PB}=\begin{bmatrix}1\\-1\\2\end{bmatrix},\quad \overrightarrow{PC}=\begin{bmatrix}3\\-2\\-1\end{bmatrix}.$$

$$\overrightarrow{PB}\times\overrightarrow{PC}=\begin{bmatrix}5\\7\\1\end{bmatrix}.$$

so

$$\overrightarrow{PA}.(\overrightarrow{PB}\times\overrightarrow{PC})=10-21+1=-10.$$

Hence the volume required is 10 units^3.

(ii) Repeat (i) with the points $P(0,0,0)$, $A(2,1,0)$, $B(1,2,0)$ and $C(3,3,2)$.

Here

$$\overrightarrow{PA}=\begin{bmatrix}2\\1\\0\end{bmatrix},\quad \overrightarrow{PB}=\begin{bmatrix}1\\2\\0\end{bmatrix},\quad \overrightarrow{PC}=\begin{bmatrix}3\\3\\2\end{bmatrix}.$$

$$\overrightarrow{PB}\times\overrightarrow{PC}=\begin{bmatrix}4\\-2\\-3\end{bmatrix},$$

so

$$\overrightarrow{PA}.(\overrightarrow{PB}\times\overrightarrow{PC})=8-2+0=6.$$

Hence the volume of the parallelepiped is 6 units^3.

Besides areas, volumes can be calculated using the cross product: in particular, volumes of parallelepipeds. A parallelepiped is a solid figure with six faces such that opposite faces are congruent parallelograms.

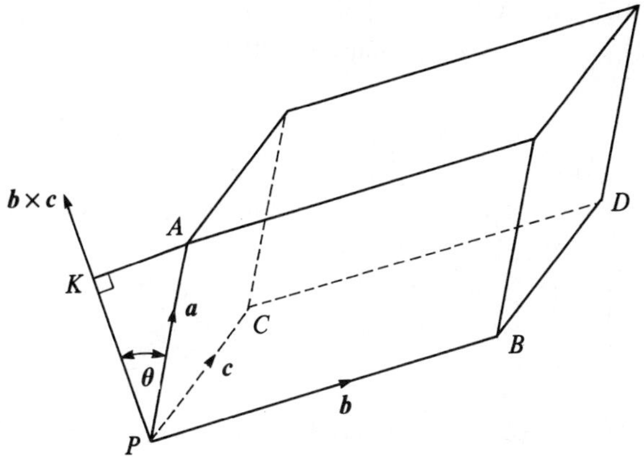

The volume of such a solid is equal to the area of the base multiplied by the height. Take $PBDC$ as the base, draw a line through P perpendicular to the base and let K be the foot of the perpendicular from A to this line. Then $|PK|$ is the height of the parallelepiped. The area of $PBDC$ is equal to $|\boldsymbol{b}\times\boldsymbol{c}|$, and the height $|PK|$ is equal to $|PA|\cos A\hat{P}K$, i.e. $|\boldsymbol{a}|\cos A\hat{P}K$. In our diagram, $A\hat{P}K$ is the angle between $\boldsymbol{a}$ and $(\boldsymbol{b}\times\boldsymbol{c})$, so

$$\boldsymbol{a}\,.\,(\boldsymbol{b}\times\boldsymbol{c})=|\boldsymbol{a}||\boldsymbol{b}\times\boldsymbol{c}|\cos A\hat{P}K=\text{volume of parallelepiped.}$$

It may happen that the direction of $\boldsymbol{b}\times\boldsymbol{c}$ is opposite to the direction of PK, in which case the angle between $\boldsymbol{a}$ and $(\boldsymbol{b}\times\boldsymbol{c})$ is $\pi-A\hat{P}K$, and

$$\begin{aligned}\boldsymbol{a}\,.\,(\boldsymbol{b}\times\boldsymbol{c})&=|\boldsymbol{a}||\boldsymbol{b}\times\boldsymbol{c}|\cos(\pi-A\hat{P}K)\\&=|\boldsymbol{a}||\boldsymbol{b}\times\boldsymbol{c}|(-\cos A\hat{P}K),\end{aligned}$$

which is the negative of what we obtained above for the other case. A volume is normally taken as a positive number, so we may combine both cases in the result

$$\text{Volume of the parallelepiped}=|\boldsymbol{a}\,.\,(\boldsymbol{b}\times\boldsymbol{c})|.$$

Notice that $\boldsymbol{a}\,.\,(\boldsymbol{b}\times\boldsymbol{c})$ is a *number*, since it is the dot product of two vectors. This form of product of three vectors is quite important, and it has a name. It is called the *scalar triple product* of $\boldsymbol{a}$, $\boldsymbol{b}$ and $\boldsymbol{c}$. The appearance of $\boldsymbol{a}\,.\,(\boldsymbol{b}\times\boldsymbol{c})$ in the formula for the volume of a parallelepiped enables us to see an unexpected property. Because the parallelepiped is the same no matter what order the three vectors are taken, we have

$$\begin{aligned}|\boldsymbol{a}\,.\,(\boldsymbol{b}\times\boldsymbol{c})|&=|\boldsymbol{b}\,.\,(\boldsymbol{c}\times\boldsymbol{a})|=|\boldsymbol{c}\,.\,(\boldsymbol{a}\times\boldsymbol{b})|=|\boldsymbol{a}\,.\,(\boldsymbol{c}\times\boldsymbol{b})|\\&=|\boldsymbol{b}\,.\,(\boldsymbol{a}\times\boldsymbol{c})|=|\boldsymbol{c}\,.\,(\boldsymbol{b}\times\boldsymbol{a})|.\end{aligned}$$

11.11 Proof that $\det A = \boldsymbol{a} . (\boldsymbol{b} \times \boldsymbol{c})$, where A is the 3×3 matrix having the vectors $\boldsymbol{a}$, $\boldsymbol{b}$ and $\boldsymbol{c}$ as its columns. Let

$$\boldsymbol{a} = \begin{bmatrix} a_1 \\ a_2 \\ a_3 \end{bmatrix}, \quad \boldsymbol{b} = \begin{bmatrix} b_1 \\ b_2 \\ b_3 \end{bmatrix}, \quad \boldsymbol{c} = \begin{bmatrix} c_1 \\ c_2 \\ c_3 \end{bmatrix}.$$

Then, expanding $\det A$ by the first column, we obtain

$$\begin{aligned} \det A &= a_1 \begin{vmatrix} b_2 & c_2 \\ b_3 & c_3 \end{vmatrix} - a_2 \begin{vmatrix} b_1 & c_1 \\ b_3 & c_3 \end{vmatrix} + a_3 \begin{vmatrix} b_1 & c_1 \\ b_2 & c_2 \end{vmatrix} \\ &= a_1(b_2c_3 - b_3c_2) - a_2(b_1c_3 - b_3c_1) + a_3(b_1c_2 - b_2c_1) \\ &= a_1(b_2c_3 - b_3c_2) + a_2(b_3c_1 - b_1c_3) + a_3(b_1c_2 - b_2c_1) \\ &= \begin{bmatrix} a_1 \\ a_2 \\ a_3 \end{bmatrix} \cdot \begin{bmatrix} b_2c_3 - b_3c_2 \\ b_3c_1 - b_1c_3 \\ b_1c_2 - b_2c_1 \end{bmatrix} \\ &= \boldsymbol{a} . (\boldsymbol{b} \times \boldsymbol{c}). \end{aligned}$$

11.12 Find whether the points O, A, B and C are coplanar, where A is $(1, 3, 0)$, B is $(0, 1, 1)$ and C is $(-2, 1, 7)$.

They are coplanar if and only if the three vectors $\overrightarrow{OA}$, $\overrightarrow{OB}$ and $\overrightarrow{OC}$ are coplanar. We therefore have to test whether these three vectors are linearly dependent. Use the standard GE process.

$$\overrightarrow{OA} = \begin{bmatrix} 1 \\ 3 \\ 0 \end{bmatrix}, \quad \overrightarrow{OB} = \begin{bmatrix} 0 \\ 1 \\ 1 \end{bmatrix}, \quad \overrightarrow{OC} = \begin{bmatrix} -2 \\ 1 \\ 7 \end{bmatrix}.$$

$$\begin{bmatrix} 1 & 0 & -2 \\ 3 & 1 & 1 \\ 0 & 1 & 7 \end{bmatrix} \to \begin{bmatrix} 1 & 0 & -2 \\ 0 & 1 & 7 \\ 0 & 1 & 7 \end{bmatrix} \to \begin{bmatrix} 1 & 0 & -2 \\ 0 & 1 & 7 \\ 0 & 0 & 0 \end{bmatrix}.$$

Hence the three vectors form a linearly dependent list (see Chapter 6), and so the four points are coplanar.

11.13 Find whether the points $P(-1, 1, -1)$, $A(2, 3, 0)$, $B(0, 1, 1)$ and $C(-2, 2, 2)$ are coplanar.

They are coplanar if and only if the vectors

$$\overrightarrow{PA} = \begin{bmatrix} 3 \\ 2 \\ 1 \end{bmatrix}, \quad \overrightarrow{PB} = \begin{bmatrix} 1 \\ 0 \\ 2 \end{bmatrix}, \quad \overrightarrow{PC} = \begin{bmatrix} -1 \\ 1 \\ 3 \end{bmatrix}$$

form a linearly dependent list. The GE process yields

$$\begin{bmatrix} 3 & 1 & -1 \\ 2 & 0 & 1 \\ 1 & 2 & 3 \end{bmatrix} \to \begin{bmatrix} 1 & \frac{1}{3} & -\frac{1}{3} \\ 2 & 0 & 1 \\ 1 & 2 & 3 \end{bmatrix} \to \begin{bmatrix} 1 & \frac{1}{3} & -\frac{1}{3} \\ 0 & -\frac{2}{3} & \frac{5}{3} \\ 0 & \frac{5}{3} & \frac{10}{3} \end{bmatrix}$$

$$\to \begin{bmatrix} 1 & \frac{1}{3} & -\frac{1}{3} \\ 0 & 1 & -\frac{5}{2} \\ 0 & \frac{5}{3} & \frac{10}{3} \end{bmatrix} \to \begin{bmatrix} 1 & \frac{1}{3} & -\frac{1}{3} \\ 0 & 1 & -\frac{5}{2} \\ 0 & 0 & 1 \end{bmatrix}.$$

Hence the three vectors form a linearly *in*dependent list, and so the given points are *not* coplanar.

These six products, however, do not all have the same value. Three of them have one value and the other three have the negative of that value. As an exercise, find out how they are grouped in this way.

We end with a further link-up between geometry and algebra. Given three 3-vectors $\boldsymbol{a}$, $\boldsymbol{b}$ and $\boldsymbol{c}$ we can construct line segments OA, OB and OC respectively representing them. We say that the vectors are *coplanar* if the three lines OA, OB and OC lie in a single plane.

Rule

Three vectors $\boldsymbol{a}$, $\boldsymbol{b}$ and $\boldsymbol{c}$ in three dimensions are coplanar if and only if $\boldsymbol{a}\,.\,(\boldsymbol{b}\times\boldsymbol{c})=0$.

To see this we need consider only the parallelepiped with one vertex at O and with A, B and C as the vertices adjacent to O. The vectors are coplanar if and only if the volume of this parallelepiped is zero (i.e. the parallelepiped is squashed flat).

Recall that three vectors $\boldsymbol{a}$, $\boldsymbol{b}$ and $\boldsymbol{c}$ form a linearly dependent list if there exist numbers l, m and n, not all zero, such that

$$l\boldsymbol{a}+m\boldsymbol{b}+n\boldsymbol{c}=\boldsymbol{0}.$$

Recall also the Equivalence Theorem, part of which stated (in the case of a 3×3 matrix A): the columns of A form a linearly independent list if and only if $\det A\neq 0$. This is logically equivalent to: the columns of A form a linearly dependent list if and only if $\det A=0$. To make the connection between this and the ideas of coplanarity and scalar triple product, consider three 3-vectors $\boldsymbol{a}$, $\boldsymbol{b}$ and $\boldsymbol{c}$. Let A be the matrix with these vectors as its columns. Then

$$\det A=\boldsymbol{a}\,.\,(\boldsymbol{b}\times\boldsymbol{c}).$$

To see this, it is necessary only to evaluate both sides. This is done in Example 11.11.

We can now see that the conditions

$$\boldsymbol{a}\,.\,(\boldsymbol{b}\times\boldsymbol{c})=0 \quad \text{and} \quad \det A=0$$

are *the same*. We therefore have:

Rule

Three vectors $\boldsymbol{a}$, $\boldsymbol{b}$ and $\boldsymbol{c}$ in three dimensions are coplanar if and only if they form a linearly dependent list.

See Examples 11.12 and 11.13 for applications of this.

As a final remark, let us note that the rule $\boldsymbol{a}\,.\,(\boldsymbol{b}\times\boldsymbol{c})=\det A$ can be a convenient way of evaluating scalar triple products and volumes of parallelepipeds.

Summary
The cross product of two 3-vectors is defined, and algebraic and geometrical properties are derived. The use of cross products in finding areas and volumes is discussed, leading to the idea of scalar triple product. The equivalence between coplanarity and linear dependence is established using the link between the scalar triple product and determinants.

Exercises

1. Evaluate the cross product of each of the following pairs of vectors (in the order given).

   (i) $\begin{bmatrix} 3 \\ 1 \\ 2 \end{bmatrix}, \begin{bmatrix} -1 \\ 1 \\ -1 \end{bmatrix}$. (ii) $\begin{bmatrix} 2 \\ 0 \\ -1 \end{bmatrix}, \begin{bmatrix} 1 \\ 3 \\ 3 \end{bmatrix}$.

   (iii) $\begin{bmatrix} 0 \\ 1 \\ -2 \end{bmatrix}, \begin{bmatrix} 3 \\ -1 \\ -2 \end{bmatrix}$. (iv) $\begin{bmatrix} -1 \\ 1 \\ -1 \end{bmatrix}, \begin{bmatrix} 3 \\ 1 \\ 2 \end{bmatrix}$.

   (v) $\begin{bmatrix} -4 \\ 4 \\ -4 \end{bmatrix}, \begin{bmatrix} 3 \\ 1 \\ 2 \end{bmatrix}$. (vi) $\begin{bmatrix} -1 \\ 2 \\ -1 \end{bmatrix}, \begin{bmatrix} 1 \\ 1 \\ 1 \end{bmatrix}$.

2. Write down the areas of the triangles OAB, where A and B are points with the pairs of vectors given in Exercise 1 as their position vectors.

3. Find the area of the triangle ABC in each case below, where A, B and C have the coordinates given.

   (i) $A(2, 1, 3)$, $B(1, 1, 0)$, $C(0, 2, 2)$.

   (ii) $A(-1, 2, -2)$, $B(-3, 0, 1)$, $C(0, 1, 0)$.

4. For each of the following pairs of planes, find a vector in the direction of the line of intersection.

   (i) $x - y + 3z - 4 = 0$.
   $2x + y - z + 5 = 0$.

   (ii) $3x + y - 2 = 0$.
   $x - 3y + z + 1 = 0$.

5. Find an equation for the plane containing the three points A, B and C, where A, B and C have the given coordinates.

   (i) $A(2, 1, 1)$, $B(4, 0, -2)$, $C(1, 1, 1)$.

   (ii) $A(0, 1, 1)$, $B(1, 0, 1)$, $C(1, 1, 0)$.

   (iii) $A(-1, 1, 2)$, $B(0, 0, 0)$, $C(3, 2, -1)$.

6. Find the volume of the parallelepiped which has $P(-1, 1, 2)$ as one vertex and $A(1, 1, -1)$, $B(0, 0, 0)$ and $C(1, 2, 0)$ as the vertices adjacent to P.

7. Repeat the calculation of Exercise 6, where P is the point $(-1,0,0)$, A is the point $(1,0,-1)$, B is the point $(1,0,1)$ and C is the point $(0,1,0)$.
8. Using the scalar triple product, find whether the given four points are coplanar.

   (i) $O(0,0,0)$, $A(1,3,5)$, $B(1,2,3)$, $C(1,0,-1)$.

   (ii) $O(0,0,0)$, $A(2,1,1)$, $B(1,1,2)$, $C(-1,2,7)$.

   (iii) $O(0,0,0)$, $A(1,-1,1)$, $B(2,2,-1)$, $C(-1,1,3)$.

   (iv) $P(2,-1,0)$, $A(3,2,2)$, $B(2,-2,-1)$, $C(4,0,-1)$.

   (v) $P(1,2,3)$, $A(-1,3,4)$, $B(3,4,7)$, $C(4,0,4)$.

ANSWERS TO EXERCISES

Chapter 1

(i) $x=1$, $y=-1$. (ii) $x=2$, $y=-3$.
(iii) $x_1=3$, $x_2=-2$, $x_3=1$. (iv) $x_1=2$, $x_2=1$, $x_3=-1$.
(v) $x_1=2$, $x_2=1$, $x_3=2$. (vi) $x_1=1$, $x_2=0$, $x_3=1$.
(vii) $x_1=1$, $x_2=-1$, $x_3=3$. (viii) $x_1=0$, $x_2=0$, $x_3=0$.
(ix) $x_1=3$, $x_2=1$, $x_3=0$. (x) $x_1=-1$, $x_2=0$, $x_3=0$.
(xi) $x_1=2$, $x_2=1$, $x_3=0$, $x_4=-1$.
(xii) $x_1=2$, $x_2=2$, $x_3=1$, $x_4=-3$.

Chapter 2

2. (i) $x=2+3t$, $y=t \quad (t\in\mathbb{R})$.
(ii) $x=-\frac{1}{2}-\frac{3}{2}t$, $y=t \quad (t\in\mathbb{R})$.
(iii) $x_1=4+t$, $x_2=1-2t$, $x_3=t \quad (t\in\mathbb{R})$.
(iv) $x_1=1-2t$, $x_2=-1+t$, $x_3=t \quad (t\in\mathbb{R})$.
(v) $x_1=1+t$, $x_2=t$, $x_3=1 \quad (t\in\mathbb{R})$.
(vi) $x_1=-2-3t$, $x_2=2+t$, $x_3=t \quad (t\in\mathbb{R})$.

3. (i) $x=2$, $y=1$. (ii) $x=3$, $y=2$.
(iii) $x_1=-1$, $x_2=1$, $x_3=2$.
(iv) $x_1=2$, $x_2=-1$, $x_3=0$.
(v) $x_1=-2$, $x_2=1$, $x_3=1$.
(vi) $x_1=4$, $x_2=3$, $x_3=2$, $x_4=1$.

4. (i) Unique solution. (ii) Infinitely many solutions.
(iii) Inconsistent. (iv) Inconsistent.
(v) Infinitely many solutions.

5. (i) $c=1$: infinitely many solutions. $c\neq 1$: inconsistent.
(ii) Consistent only for $k=6$.

6. (i) $c=-2$: no solution. $c=2$: infinitely many solutions. Otherwise: unique solution.
(ii) $c=\pm\sqrt{3}$: no solution. Otherwise: unique solution.

(iii) $c=-3$: no solution. $c=3$: infinitely many solutions. Otherwise: unique solution.

(iv) $c=0$, $c=\sqrt{6}$, $c=-\sqrt{6}$: infinitely many solutions. Otherwise: unique solution.

Chapter 3

2. (i) $\begin{bmatrix}-1\\-2\\2\end{bmatrix}$. (ii) $\begin{bmatrix}4\\-1\\3\end{bmatrix}$. (iii) $\begin{bmatrix}-3\\-3\end{bmatrix}$. (iv) $\begin{bmatrix}4\\-1\\3\\3\\0\end{bmatrix}$.

(v) $\begin{bmatrix}15\\15\\15\end{bmatrix}$. (vi) $\begin{bmatrix}-2\\1\\3\end{bmatrix}$.

3. $AB=\begin{bmatrix}19 & -1 & 6 & 13\\16 & -8 & -8 & 6\end{bmatrix}$, $AD=\begin{bmatrix}21 & -7\\13 & -16\end{bmatrix}$,

$BC=\begin{bmatrix}0 & 3 & 7\\8 & -4 & 6\\9 & 3 & 3\end{bmatrix}$, $CB=\begin{bmatrix}8 & -1 & -1 & 6\\4 & -8 & -4 & -6\\9 & 0 & -3 & 9\\4 & 0 & 4 & 2\end{bmatrix}$,

$CD=\begin{bmatrix}8 & -5\\1 & -8\\9 & -6\\5 & 0\end{bmatrix}$.

DA also exists. $A(BC)$ of course should be the same as $(AB)C$.

4. (i) $\begin{bmatrix}11 & 10\\2 & -2\\19 & 16\end{bmatrix}$. (ii) $\begin{bmatrix}3 & 4 & 2\\-2 & -2 & 10\\7 & 4 & 3\\7 & 8 & 4\end{bmatrix}$.

(iii) $\begin{bmatrix}3 & 4\\1 & 2\end{bmatrix}$. (iv) $\begin{bmatrix}2\\-6\end{bmatrix}$.

5. $\begin{bmatrix}8 & 8 & 13\\8 & 5 & 8\\13 & 8 & 8\end{bmatrix}$.

6. A must be $p\times q$ and B must be $q\times p$, for some numbers p and q.

Chapter 4

1. $A^2=\begin{bmatrix}1 & 2 & 3\\0 & 1 & 2\\0 & 0 & 1\end{bmatrix}$, $A^3=\begin{bmatrix}1 & 3 & 6\\0 & 1 & 3\\0 & 0 & 1\end{bmatrix}$,

$$A^4=\begin{bmatrix}1&4&10\\0&1&4\\0&0&1\end{bmatrix},\quad B^2=\begin{bmatrix}1&0&0\\2&1&0\\3&2&1\end{bmatrix},$$

$$B^3=\begin{bmatrix}1&0&0\\3&1&0\\6&3&1\end{bmatrix},\quad B^4=\begin{bmatrix}1&0&0\\4&1&0\\10&4&1\end{bmatrix}.$$

3. (i) $\begin{bmatrix}0&-1&5\\0&4&-3\\0&0&-1\end{bmatrix}$. (ii) $\begin{bmatrix}x_1+2x_2+x_3\\x_2-2x_3\\x_3\end{bmatrix}$.

(iii) $\begin{bmatrix}-1&0&0\\-3&-2&0\\9&5&3\end{bmatrix}$. (iv) $[6\quad 5\quad 3]$.

4. $x_1=2,\ x_2=6,\ x_3=1$.

6. Symmetric:

$$\begin{bmatrix}1&2\\2&3\end{bmatrix},\ \begin{bmatrix}1&0\\0&-1\end{bmatrix},\ \begin{bmatrix}2&3&1\\3&0&-1\\1&-1&2\end{bmatrix},$$

$$\begin{bmatrix}1&0&1\\0&1&0\\1&0&1\end{bmatrix},\ \begin{bmatrix}1&1&0\\1&0&-1\\0&-1&-1\end{bmatrix}.$$

Skew-symmetric:

$$\begin{bmatrix}0&2\\-2&0\end{bmatrix},\ \begin{bmatrix}0&1&-2\\-1&0&3\\2&-3&0\end{bmatrix}.$$

9. (i) Interchange rows 2 and 3.
(ii) Add twice row 3 to row 1.
(iii) Subtract three times row 1 from row 3.
(iv) Multiply row 2 by -2.

10. $$T=\begin{bmatrix}1&0&0\\0&1&0\\0&0&\frac{1}{6}\end{bmatrix}\begin{bmatrix}1&0&0\\0&1&0\\0&1&1\end{bmatrix}\begin{bmatrix}1&0&0\\0&1&0\\-2&0&1\end{bmatrix}\begin{bmatrix}0&1&0\\1&0&0\\0&0&1\end{bmatrix},$$

and

$$TA=\begin{bmatrix}1&2&-1\\0&1&3\\0&0&1\end{bmatrix}.$$

11. $$T=\begin{bmatrix}1&0&0\\0&1&0\\0&0&-\frac{1}{6}\end{bmatrix}\begin{bmatrix}1&0&0\\0&1&0\\0&-3&1\end{bmatrix}\begin{bmatrix}1&0&0\\0&\frac{1}{2}&0\\0&0&1\end{bmatrix}$$

$$\times\begin{bmatrix}1&0&0\\0&1&0\\-2&0&1\end{bmatrix}\begin{bmatrix}1&0&0\\1&1&0\\0&0&1\end{bmatrix}.$$

and

$$TA=\begin{bmatrix}1&-1&2&1\\0&1&1&1\\0&0&1&1\end{bmatrix}.$$

Chapter 5

1. Invertible: (i), (ii), (iv), (v), (vi), (viii), (xi).
2. $\begin{bmatrix} 1/a & 0 & 0 \\ 0 & 1/b & 0 \\ 0 & 0 & 1/c \end{bmatrix}$.

Chapter 6

1. $x=-1$, $y=2$.
2. (i) Yes. (ii) No. (iii) Yes. (iv) No.
3. In each case we give the coefficients in a linear combination which is equal to **0** (the vectors taken in the order given).
   (i) 2, −1. (ii) 6, −7, 2. (iii) 4, −1, 3.
   (iv) −9, 7, 16. (v) 5, −4, 3. (vi) 1, −13, 5.
4. LD: (ii), (iii), (vi).
   LI: (i), (iv), (v), (vii).
5. 1, 2, 2, 1, 2, 2, 3, 3, 2, 1, 1, 2, 1, 2, 3, 4, 4, 2, 3.
6. Rank of $\boldsymbol{x}\boldsymbol{y}^{\mathrm{T}}$ is 1. This holds for any p.

Chapter 7

1. −7, −2, 8, 0, −28, −7, 7, −3.
2. 2, −4, −9, 0, −10, 10, 0, 0, −3.
3. −16, −11.
4. No, only for $p \times p$ skew-symmetric matrices with *odd* p.

8. $\det A=2$. $\operatorname{adj} A = \begin{bmatrix} -1 & 1 & 1 \\ 1 & -1 & 1 \\ 1 & 1 & -1 \end{bmatrix}$.

Chapter 8

(i) Ranks 2, 2; unique solution.
(ii) Ranks 1, 2; inconsistent.
(iii) Ranks 1, 1; infinitely many solutions.
(iv) Ranks 2, 2; unique solution.
(v) Ranks 3, 3; unique solution.
(vi) Ranks 3, 3; unique solution.
(vii) Ranks 2, 3; inconsistent.
(viii) Ranks 2, 2; infinitely many solutions.
(ix) Ranks 2, 2; infinitely many solutions.
(x) Ranks 2, 2; infinitely many solutions.
(xi) Ranks 2, 2; infinitely many solutions.
(xii) Ranks 3, 3; unique solution.

Chapter 9

1. C: $\boldsymbol{b}-\boldsymbol{a}$, D: $-\boldsymbol{a}$, E: $-\boldsymbol{b}$, F: $\boldsymbol{a}-\boldsymbol{b}$.

5. (i) $\begin{bmatrix}2\\-1\\3\end{bmatrix}$. (ii) $\begin{bmatrix}-2\\1\\-3\end{bmatrix}$. (iii) $\begin{bmatrix}-2\\-2\\-2\end{bmatrix}$. (iv) $\begin{bmatrix}0\\-2\\1\end{bmatrix}$.

(v) $\begin{bmatrix}1\\0\\-1\end{bmatrix}$.

6. (i) $\begin{bmatrix}0\\1\\4\end{bmatrix}$. (ii) $\begin{bmatrix}8\\5\\3\end{bmatrix}$. (iii) $\begin{bmatrix}6\\6\\6\end{bmatrix}$. (iv) $\begin{bmatrix}12\\-10\\16\end{bmatrix}$.

(v) $\begin{bmatrix}10\\5\\-5\end{bmatrix}$.

9. (i) $\begin{bmatrix}\frac{1}{\sqrt{2}}\\0\\-\frac{1}{\sqrt{2}}\end{bmatrix}$. (ii) $\begin{bmatrix}\frac{2}{3}\\\frac{2}{3}\\\frac{1}{3}\end{bmatrix}$. (iii) $\begin{bmatrix}\frac{1}{\sqrt{6}}\\-\frac{2}{\sqrt{6}}\\-\frac{1}{\sqrt{6}}\end{bmatrix}$. (iv) $\begin{bmatrix}\frac{1}{\sqrt{3}}\\\frac{1}{\sqrt{3}}\\\frac{1}{\sqrt{3}}\end{bmatrix}$.

Chapter 10

1. (i) $x=t,\ y=1-t,\ z=3-2t\quad (t\in\mathbb{R})$.
(ii) $x=1,\ y=1-t,\ z=-2+2t\quad (t\in\mathbb{R})$.
(iii) $x=-1,\ y=2,\ z=4-11t\quad (t\in\mathbb{R})$.
(iv) $x=1+t,\ y=1+t,\ z=1+t\quad (t\in\mathbb{R})$.
(v) $x=3t,\ y=-t,\ z=2t\quad (t\in\mathbb{R})$.

2. (i) $\begin{bmatrix}-1\\2\\-5\end{bmatrix}$. (ii) $\begin{bmatrix}2\\-1\\1\end{bmatrix}$. (iii) $\begin{bmatrix}-3\\0\\-1\end{bmatrix}$.

3. (i) Intersect at (0, 1, 1).
(ii) Intersect at (0, 3, 1).
(iii) Do not intersect.
(iv) These two sets of equations represent the same line.

4. (i) $\dfrac{2}{\sqrt{6}}$. (ii) $-\dfrac{1}{6}$. (iii) $\dfrac{2}{\sqrt{42}}$. (iv) $-\dfrac{2}{\sqrt{6}}$.

5. $\cos\hat{A}=-\dfrac{1}{\sqrt{2}}$, $\cos\hat{B}=\dfrac{5}{\sqrt{34}}$, $\cos\hat{C}=\dfrac{4}{\sqrt{17}}$.

The largest angle is $\hat{A}$, which is $3\pi/4$ radians.

6. For example,

$$\begin{bmatrix} 1 \\ -2 \\ 3 \end{bmatrix}.$$

7. 7 units.

8. (i) $\sqrt{2}$. (ii) $\sqrt{24}$. (iii) 0 (the point lies on the line).

9. (i) $x-y+2z-3=0$. (ii) $4x+5y+6z+1=0$.
(iii) $y+3z-18=0$. (iv) $x+y+z-3=0$.

10. (i) 0 (the planes are perpendicular).

(ii) $\dfrac{5}{\sqrt{180}}$. (iii) $\dfrac{1}{2}$. (iv) $\dfrac{2}{\sqrt{14}}$.

11. (i) Straight line. (ii) Single point.
(iii) Empty. (iv) Single point.

12. (i) 0. (ii) $\dfrac{6}{\sqrt{14}}$. (iii) $\sqrt{3}$. (iv) $\dfrac{6}{\sqrt{42}}$. (v) $\dfrac{1}{\sqrt{42}}$.

Chapter 11

1. (i) $\begin{bmatrix} -3 \\ 1 \\ 4 \end{bmatrix}$. (ii) $\begin{bmatrix} 3 \\ -7 \\ 6 \end{bmatrix}$. (iii) $\begin{bmatrix} -4 \\ -6 \\ -3 \end{bmatrix}$.

(iv) $\begin{bmatrix} 3 \\ -1 \\ -4 \end{bmatrix}$. (v) $\begin{bmatrix} 4 \\ 20 \\ -16 \end{bmatrix}$. (vi) $\begin{bmatrix} 3 \\ 0 \\ -3 \end{bmatrix}$.

2. (i) $\frac{1}{2}\sqrt{26}$. (ii) $\frac{1}{2}\sqrt{94}$. (iii) $\frac{1}{2}\sqrt{61}$.
(iv) $\frac{1}{2}\sqrt{26}$. (v) $\frac{1}{2}\sqrt{672}$. (vi) $\frac{1}{2}\sqrt{18}$.

3. (i) $\frac{1}{2}\sqrt{35}$. (ii) $\frac{1}{2}\sqrt{66}$.

4. (i) $\begin{bmatrix} -2 \\ 7 \\ 3 \end{bmatrix}$. (ii) $\begin{bmatrix} 1 \\ -3 \\ -10 \end{bmatrix}$.

5. (i) $3y-z-2=0$. (ii) $x+y+z-2=0$.
(iii) $x-y+z=0$.

6. 3 units^3.

7. 4 units^3.

8. (i), (ii) and (iv) are coplanar. The others are not.

SAMPLE TEST PAPERS

Paper 1

1

(i) Let X be a 3×4 matrix. What size must the matrix Y be if the product XYX is to exist? For such a matrix Y, what is the size of the matrix XYX?

Calculate AB or BA (or both, if both exist), where

$$A=\begin{bmatrix}-1 & 2\\ 0 & 1\\ 3 & -1\end{bmatrix} \quad\text{and}\quad B=\begin{bmatrix}0 & 2 & 3\\ 1 & -1 & 0\end{bmatrix}.$$

(ii) Find the values of t for which the following equations have (a) a unique solution, and (b) infinitely many solutions.

$$\begin{aligned} tx+4y&=0\\ (t-1)x+ty&=0.\end{aligned}$$

(iii) Let X and Y be $p\times p$ symmetric matrices. Is the matrix $XY-YX$ symmetric? Is it skew-symmetric? If P and Q are skew-symmetric matrices, what can be said about the symmetry or skew-symmetry of the matrix $PQ-QP$?

2

Show that the list

$$\left(\begin{bmatrix}1\\2\\1\end{bmatrix}, \begin{bmatrix}1\\4\\-1\end{bmatrix}, \begin{bmatrix}1\\-1\\a^2+3a\end{bmatrix}\right)$$

is linearly independent if and only if $a=1$ or $a=-4$. For each of these values of a, find a non-trivial linear combination of these vectors which is equal to the zero vector.

What is the rank of the matrix

$$A=\begin{bmatrix}1 & 1 & 1\\ 2 & 4 & -1\\ 1 & -1 & a^2+3a\end{bmatrix},$$

when $a=1$ or $a=-4$?

Find the inverse of A when $a=0$, and hence or otherwise solve the equation

$$Ax=\begin{bmatrix}-1\\-2\\1\end{bmatrix}$$

in the case when $a=0$.

3

(i) Show that the determinant of a 3×3 skew-symmetric matrix is equal to zero. Do all skew-symmetric matrices have determinant equal to zero? Justify your answer.
(ii) Explain what is meant by an *elementary matrix*. Give examples of the three different kinds of elementary matrix, and explain their connection with the Gaussian elimination process.
(iii) Let A, B and C be the points $(1,0,0)$, $(0,2,0)$ and $(0,0,2)$ respectively. Using the cross product of vectors, or otherwise, find the surface area of the tetrahedron $OABC$.

4

Let $A(2,1,-4)$, $B(0,-1,-6)$, $C(3,0,-1)$ and $D(-3,-4,-3)$ be four points in space. Find parametric equations for the straight lines AB and CD. Hence show that these two lines do not intersect. Let P and Q be points on AB and CD respectively such that PQ is perpendicular to both AB and CD. Calculate the length of PQ.

Find an equation for the locus of all midpoints of line segments joining a point on AB to a point on CD. Deduce that the locus is a plane.

Paper 2

1

(i) Let

$$A=\begin{bmatrix}2 & 1\\0 & 1\end{bmatrix},\quad B=\begin{bmatrix}1 & -1 & 0\\2 & 2 & 1\end{bmatrix}\quad\text{and}\quad C=\begin{bmatrix}1 & 1\\2 & 0\\-1 & -1\\2 & 3\end{bmatrix}.$$

Calculate all possible products of two of these matrices. Is it possible to multiply them all together? If so, in what order? Calculate any such product of all three.

(ii) Let

$$X=\begin{bmatrix}1 & 2 & -1\\2 & 3 & 0\\-1 & 0 & -2\end{bmatrix}.$$

Find whether X is invertible. If it is, find its inverse.

(iii) Define a *skew-symmetric* matrix. Explain why the entries on the main diagonal of a skew-symmetric matrix must all be zero. Let H be the matrix

$$\begin{bmatrix}0 & 1\\-1 & 0\end{bmatrix}.$$

Show that $H^2+I=0$, that H is invertible, and that $H^{-1}=H^{\mathrm{T}}$.

2

(i) Let A be a $p\times q$ matrix and let $\boldsymbol{b}$ be a p-vector. In the matrix equation $A\boldsymbol{x}=\boldsymbol{b}$, what condition must be satisfied by the rank of A and the rank of the augmented matrix $[A\,|\,\boldsymbol{b}]$ if the equation is to have no solutions? Prove that the following set of equations has no solutions.

$$\begin{aligned} x+2y+3z&=1\\ x+\ y+\ z&=2\\ 5x+7y+9z&=6. \end{aligned}$$

(ii) Find whether the list

$$\left(\begin{bmatrix}2\\1\\-3\end{bmatrix},\begin{bmatrix}1\\-2\\-1\end{bmatrix},\begin{bmatrix}-1\\-3\\2\end{bmatrix}\right)$$

is linearly dependent or linearly independent.

(iii) Find all values of c for which the equations

$$\begin{aligned} (c+1)x+\qquad 2y&=0\\ 3x+(c-1)y&=0 \end{aligned}$$

have a solution other than $x=y=0$.

3

(i) Evaluate the determinant

$$\begin{vmatrix}2 & 1 & 3 & -1\\-2 & 0 & 1 & 1\\1 & 0 & 2 & 2\\3 & 0 & -1 & 1\end{vmatrix}$$

(ii) Explain what is meant by the adjoint (adj A) of a square matrix A. Show that, for any 3×3 matrix A,

$$A(\text{adj } A) = (\det A)I,$$

where I is the 3×3 identity matrix.

(iii) Find an equation for the plane containing the three points $A(2, 1, 1)$, $B(-1, 5, 9)$ and $C(4, 5, -1)$.

4

Define the *dot product* $\boldsymbol{a} \cdot \boldsymbol{b}$ of two non-zero vectors $\boldsymbol{a}$ and $\boldsymbol{b}$.

(i) Let $OABC$ be a tetrahedron, O being the origin. Suppose that OC is perpendicular to AB and that OB is perpendicular to AB. Prove that OA is perpendicular to BC.

(ii) Let P and Q be the points $(1, 0, -1)$ and $(0, 1, 1)$ respectively. Find all unit vectors $\boldsymbol{u}$ which are perpendicular to $\overrightarrow{OP}$ and which make angle $\pi/3$ (60°) with $\overrightarrow{OQ}$.

Paper 3

1

(i) Let

$$A=\begin{bmatrix}1 & 0\\ 2 & -1\\ 0 & 2\\ 1 & 3\end{bmatrix},\quad B=\begin{bmatrix}1 & 1 & 0 & -1\\ -1 & 0 & 2 & 0\\ 0 & 0 & 1 & 1\end{bmatrix}\quad\text{and}\quad C=\begin{bmatrix}1 & 0 & 1\\ -1 & 1 & 0\end{bmatrix}.$$

Evaluate the product CB. Evaluate every other product of A or C with B. There are exactly three orders in which it is possible to multiply A, B and C all together. Write these down but do not evaluate the products. State the sizes of the three product matrices.

(ii) Define the *rank* of a matrix. Calculate the rank of the matrix

$$\begin{bmatrix}1 & -3 & 4\\ 2 & -1 & 7\\ 2 & 4 & 6\end{bmatrix}.$$

Use your answer in determining (without actually solving them) whether the following equations are consistent.

$$\begin{aligned}x-3y+4z&=0\\ 2x-\ y+7z&=4\\ 2x+4y+6z&=8.\end{aligned}$$

(iii) Show that the product of two upper triangular 3×3 matrices is upper triangular.

2

Solve the system of equations

$$\begin{aligned}x+2y-\ z&=\ \ 4\\ 2x-\ y+\ z&=-3\qquad (*)\\ -x+\ y+4z&=-7.\end{aligned}$$

Show that the inverse of an invertible symmetric matrix is symmetric, and verify this by finding the inverse of

$$P=\begin{bmatrix}1 & 1 & 0\\ 1 & 2 & 1\\ 0 & 1 & 0\end{bmatrix}.$$

Let A be a 3×3 matrix and let $\boldsymbol{b}$ be a 3-vector. Show that if $\boldsymbol{c}$ is a solution to the equation $A\boldsymbol{x}=\boldsymbol{b}$ then $P^{-1}\boldsymbol{c}$ is a solution to the equation $(AP)\boldsymbol{x}=\boldsymbol{b}$. Use your earlier results to find a solution to

$$(AP)\boldsymbol{x}=\begin{bmatrix}4\\ -3\\ -7\end{bmatrix},$$

where A is the matrix of coefficients on the left-hand sides of equations (*) above.

3

(i) Let

$$A = \begin{bmatrix} 1 & t & 0 \\ 1+t & 1 & 5 \\ 0 & -t & t \end{bmatrix}, \quad \text{where } t \in \mathbb{R}.$$

Evaluate det A and hence find all values of t for which A is singular.

(ii) Let X be a 3×1 matrix and let Y be a 1×3 matrix. Show that XY is a *singular* 3×3 matrix.

(iii) Let A, B, C and P be points with coordinates $(2, 1, 1)$, $(-4, -2, 1)$, $(1, 2, 3)$ and $(-1, -1, 2)$ respectively. Find which of the angles $B\hat{P}C$, $C\hat{P}A$ and $A\hat{P}B$ is the smallest.

4

Give the definition of the *cross product* $\boldsymbol{a} \times \boldsymbol{b}$ of two non-zero vectors $\boldsymbol{a}$ and $\boldsymbol{b}$.

Find an equation for the plane π through the points $A(1, 1, 4)$, $B(3, -2, 4)$ and $C(3, -1, 1)$.

What is the perpendicular distance of the point $X(1, -3, 5)$ from the plane π? Find the volume of the parallelepiped which has X as one vertex and A, B and C as the vertices adjacent to X. Find the coordinates of the vertex of this parallelepiped which is farthest from X.

INDEX

FURTHER READING

As indicated in the Preface, there are many books on linear algebra, and as suggested there, not many which contain treatments which are sympathetic with the approach taken in this book. Here is a selection which the reader may usefully refer to or take as a starting point for further study.

[1] F. Ayres, *Matrices*. Schaum's Outline Series, McGraw-Hill, 1968.

This is a book of problems and solutions.

[2] H. Anton, *Elementary Linear Algebra*, 4th edition. John Wiley, 1984.

[3] D. T. Finkbeiner, *Elements of Linear Algebra*, 3rd edition. Freeman, 1978.

[4] B. Kolman, *Elementary Linear Algebra*, 4th edition. Collier Macmillan, 1986.

[5] I. Reiner, *Introduction to Linear Algebra and Matrix Theory*. Holt, Rinehart & Winston, 1971.

These are four very similar books. They are all rather more advanced and rather more substantial than this book, but there is common material, and their contents should for the most part be accessible to the interested reader of this book.

[6] P. J. Kelly & E. G. Straus, *Elements of Analytical Geometry*. Scott Foresman, 1970.

[7] J. H. Kindle, *Plane and Solid Analytic Geometry*. Schaum's Outline Series, McGraw-Hill, 1950.

These two books, as their titles suggest, are about geometry rather than algebra, but they may be useful as background and/or further reading for the more geometrical aspects of this book.